Marine Renewables: A Wave of Power

Sana

Table of Contents

Chapter I: Introduction 1

    1.1   Research goals 4

Chapter II: Literature Review 7

    2.1   Tidal energy.. 7

       2.1.1   Resource and technology overview... 7

       2.1.2   Advantages 9

       2.1.3   Disadvantages 11

    2.2   Diesel generated electricity 12

       2.2.1   Technology overview... 13

       2.2.2   Advantages.. 13

       2.2.3   Disadvantages 14

    2.3   Stakeholder engagement... 16

    2.4   Analyzing the spatial feasibility and suitability of tidal energy................................ 19

       2.4.1   Marine Spatial Planning... 19

       2.4.2   Geographic Information System Multi-Criteria Decision Analysis 20

       2.4.3   GIS suitability mapping and GIS-MCDA mapping for tidal energy 23

    2.5   Economic feasibility 24

       2.5.1   Tidal energy costs.. 24

       2.5.2   Levelized Cost of Energy... 25

       2.5.3   Externalities and willingness to pay.. 26

Chapter III: Study area overview... 28

    3.1   Study area 28

       3.1.1   Existing MSP 28

3.1.2   Environment brief 30

3.1.3   Social brief.. 32

3.1.4   Economic brief.. 33

Chapter IV: Phase I methods.. 35

4.1   Interviews and GIS-MCDA... 35

4.1.1   The Analytic Hierarchy Process 38

4.2   Data overview 40

4.2.1   Project set up 41

4.3   Resource and marine use layers creation and normalization... 41

4.3.1   Tidal stream turbine technical feasibility and constraint layers 42

4.3.2   Tidal Resource sub-model 44

4.3.3   Economic Uses sub-model input layers.. 47

4.3.4   Social Uses sub-model input layers 49

4.3.5   Marine Conservation Value sub-model 50

4.4   Phase I Weighted Overlay sub-model and model creation 50

Chapter V: Phase I results 53

5.1   Phase I interviews 53

5.2   Phase I GIS-MCDA sensitivity analyses 62

5.2.1   Sub-model sensitivity analysis 62

5.2.2   Model sensitivity analysis 62

5.3   Phase I GIS-MCDA results 63

5.3.1   North Vancouver Island sub-region.. 63

5.3.2   Central Coast sub-region 64

5.3.3   North Coast sub-region 65

5.3.4   Haida Gwaii sub-region 66

Chapter VI: Phase II methods... 67

6.1   Including economic site considerations and constraints for a candidate site.................... 67

6.1.1   Site distance from ports 67

6.1.2   Site distance from communities... 68

6.1.3   Site depth 70

6.1.4   Site MSP zoning.. 72

6.1.5   Phase I Tidal Development Perspective suitability model 73

6.2   Phase II weighted overlay 74

6.3   Candidate community interview..74

Chapter VII: Phase II results 75

7.1   Phase II weighted overlay 75

7.1.1   Phase II mapping sensitivity analysis 76

7.2   Candidate community identification 76

7.3   Candidate community overview 78

7.3.1   Candidate tidal site overview 81

7.4   Phase II interview results 83

Chapter VIII: Phase III methods..86

8.1   Diesel LCOE analysis...86

8.2   Externality included LCOE analysis..87

8.3   Tidal LCOE analysis 88

8.3.1   Project specifications..89

8.3.2   Turbine specifications...89

8.3.3   Site specifications 91

Chapter IX: Phase III results 92

9.1   Existing diesel infrastructure LCOE 92

9.2 CO_2 equivalent emissions externality LCOE 92

9.3   Skidegate Inlet tidal site LCOE 93

Chapter X: Discussion..95

10.1 Electricity in remote coastal diesel reliant First Nations communities...........................95

10.1.1   Barriers to tidal energy electrification and pathways forward 96

10.2   Integrating tidal energy into MaPP region..98

10.2.1   Incorporating marine uses into tidal energy suitability analyses.........................98

10.2.2   Future planning for tidal energy in the MaPP region 99

10.3   Candidate community case study 101

10.3.1   MSP suitability of the identified site...101

10.3.2   Economic viability of the Skidegate Inlet site 102

10.3.3   TSTs and Haida Gwaii's energy future 103

10.4   Levelized Cost of Energy shortcomings...105

10.4.1   Levelized cost of energy analysis shortcomings 105

10.4.2   Natural Resources Canada Tidal Project Cost Estimation Tool 106

10.5   Developing and supporting the tidal energy industry in BC 108

10.5.1   Opportunities for community self-sufficiency through tidal energy 108

10.6   Limitations 110

10.6.1   Data, processing, and methods 110

10.6.2   Stakeholder engagement 112

10.6.3   Scope and scale of the research and interdisciplinary methods 114

Chapter XI: Conclusions 116

11.1   Summary of findings 116

11.2   Future research recommendations 117

References 119

Appendix A: Human Research Ethics Certification 147

Appendix B: Analytic Hierarchy Process Constraints, Criteria, and Sensitivity Analysis Methods 151

Appendix C: Datasets and Sources 155

Appendix D: Diesel Fuel Price for Current Suitability and Tidal Stream Turbine Device Specifications ... 160

Appendix E: Layer Histograms 162

Appendix F: Economic, Social, and Environmental Use Layers 164

Appendix G: AHP Matrices and Sub-model Sensitivity Analyses 169

Appendix H: Phase I Model Sensitivity Analysis 172

Appendix I: Phase I GIS-MCDA Sub-model Results 174

Appendix J: Port Selection Rationale 178

Appendix K: Phase II Equal Weights Model Results 179

Appendix L: Tidal Stream Turbine Levelized Cost of Energy Inputs 180

Chapter I: Introduction

With the latest Intergovernmental Panel on Climate Change report calling for *"rapid and far-reaching transitions in energy, land, urban and infrastructure, and industrial systems"* to avoid the societal and environmental impacts associated with exceeding the 1.5°C warming threshold, our obligation to mitigate climate change has never been so imperative (Allen et al., 2018, p. 17). A subsequent United Nations Environmental Programme report in 2019 illustrated bleak findings, as countries collectively failed to stop the growth of Greenhouse Gas (GHG) emissions and thus more significant and rapid reductions are needed by the global community (UNEP, 2019). Fossil fuel-based electricity and heat generation account for 25% of global greenhouse gas (GHG) emissions and thus the transition to alternative energy sources is a crucial target for reductions (Jenniches, 2018). Renewable energy technologies (RET) are prime candidates to mitigate climate change while also providing energy security, promoting economic growth, developing new industries, creating jobs, and diversifying electricity production (O'Rourke et al., 2009; Ruano-Chamorro et al., 2018).

Over the past 30 years, RETs such as solar and wind energy have seen substantial technological improvements allowing them to become alternatives to fossil-fuel electricity production (IRENA, 2019; O'Rourke et al., 2009). Despite advancements, RETs are still hampered by cost, resource availability, and unpredictability, as well as frequent spatial conflicts with anthropogenic and ecological land uses (Barrington-Leigh & Ouliaris, 2017; Dijkman & Benders, 2010; Sen & Ganguly, 2017). Marine Renewable Energy (MRE) technologies specifically tidal stream (also referred to as tidal energy in this paper), wave, and offshore wind can provide the same benefits as their onshore counterparts coupled with greater predictability and minimal terrestrial footprints (Bedard et al., 2010; Borthwick, 2016; Quero García et al., 2019; Robertson et al., 2017; O'Rourke et al., 2009).

Tidal energy has several advantages; namely its predictability, high-capacity factor, minimal environmental impact, and high resource potential in proximity to many coastal communities (Bedard et al., 2010; Bonar et al., 2015; Copping & Hemery, 2020; OES, 2020; Segura et al., 2017b). Despite these advantages, the nascent industry is associated with considerable innovation costs, substantial device capital costs, and high investment uncertainty, along with issues such as delineating areas for development and the inclusion of stakeholders in the decision-making process (Jahanshahi et al., 2019; Jenkins et al., 2018; MacDougall, 2017; Sangiuliano, 2017b; Segura et al., 2017b; Segura et al., 2018; Vazquez & Iglesias, 2015).

Tidal energy has also been identified as a promising technology for smaller scale development in remote coastal communities, such as those in the province of British Columbia (BC) which has over 2 GW of estimated tidal power potential (De Groot & Bailey, 2016; Roy et al., 2018; Segura et al., 2018; Tawil et al., 2018; Triton Consultants ltd., 2002). Although BC relies on large scale hydroelectric dams for over 95% of its energy, remote communities not connected to the grid produce 0.5% of BCs total energy from diesel generators (Government of Canada, 2016; MEMPR, 2020; NRCan, 2011). A remote off grid community is defined as a permanent settlement (longer than five years) with more than 10 dwellings that is currently not connected to the North American electrical grid, or the natural gas network (NRCan, 2011). There are 86 remote communities reliant on diesel in BC, 18 of which are predominantly First Nations in composition and coastal, thus tidal may be a suitable RET (NRCan, 2011). Diesel reliance yields a suite of disadvantages including higher electricity costs, health impacts, environmental degradation, community growth constraints, and the increased likelihood of blackouts and brownouts (Karanasios & Parker, 2018; Kennedy, 2018; NRCan, 2011; NRCan, 2013; Price Waterhouse Cooper, 2015; Rezaei & Dowlatabadi, 2016). Several of these communities are located near tidal resources, and thus tidal energy may be an apt alternative to diesel (NRCan, 2011; Triton Consultants Ltd., 2002). With policy targets to reduce diesel reliance by 80% in BCs remote communities by 2030, assessing tidal energy is now paramount to aid the examination and determination of the most suitable renewable resources for these communities (Government of British Columbia, 2018).

Although tidal energy avoids spatial competition with uses on land, it requires access to areas of the marine space and thus must be balanced with existing uses and Marine Spatial Planning (MSP) priorities (Kerr et al., 2014a; Wright, 2015). These are diverse and numerous including, but not limited to; environment (e.g. ecological uses, sensitive habitats, species distributions), economic (e.g. fishing, shipping, transportation, dredging), and social (e.g. recreation, cultural sites, visual amenity) uses. Geographic Information Systems (GIS) and Multi-Criteria Decision Analysis (MCDA) represent a cost-effective way of delineating existing uses and investigating how emerging uses such as tidal energy can be integrated within the marine space (Davies et al., 2014; Gimpel et al., 2015; Kim et al., 2012).

Stakeholder engagement represents a crucial means of incorporating social values when assessing locations for tidal energy (Gopnik et al., 2012; Janssen et al., 2014; Shucksmith & Kelly, 2014). Engagement allows for local communities and stakeholders to contribute to early-stage planning, ensuring that development and climate action align with communal values and are integrated in ways which enhance the lives of community members. Additionally, early, and ongoing engagement reduces the likelihood of opposition later in the process when it may be more costly (Dalton et al., 2015; Frazão Santos et al., 2018; Kerr et al., 2014b; Richardson, 2018;

Ruano-Chamorro et al., 2018). It is also necessary to understand whether tidal energy is viewed as a culturally appropriate resource, along with gaining an understanding of the communities' visions for their energy projects and to determine whether the broader literature on remote electrification of communities applies in terms of government and academic rationale versus community views (Rezaei & Dowlatabadi, 2016).

While stakeholder support and suitable locations for development are crucial, cost is arguably the main barrier to the development of the industry and constitutes a substantial tidal disadvantage compared to more developed RETs (Vazquez & Iglesias, 2016b). Furthermore, the quantification and incorporation of externalities associated with tidal and diesel is vital to truly assess and compare the costs and benefits of each (Eidelwein et al., 2018; Lehmann et al., 2018).

1.1   Research goals

This research seeks to contribute to global GHG emissions reduction efforts while ensuring environmental protection, minimizing infringement upon existing uses, and enhancing the quality of life in First Nations remote coastal diesel reliant communities (RCDRC) in BC by investigating the spatial suitability, biophysical capability, and economic feasibility of tidal energy using an integrated interdisciplinary framework. This analysis seeks to answer the following questions:

1. What are the potential benefits and challenges of replacing existing diesel generators with tidal energy in remote off grid communities in the Marine Plan Partnership for the North Pacific Coast (MaPP) region?

2. In accordance with local stakeholder input, how should suitability mapping inform the development of tidal energy and how can tidal devices be integrated

within existing electricity generation systems to enhance the standard of living in a candidate community?

3. Using GIS MCDA suitability mapping, can tidal energy be sustainably (from environment, social, and economic perspectives) integrated within the broader context of human and natural uses of the marine space in BC?

4. What is the existing Levelized Cost of Energy (LCOE) for diesel and an externality included diesel LCOE within a candidate community that tidal would have to compete with, and what is the LCOE for a potential tidal site?

5. What are the benefits and drawbacks associated with assessing renewable energy development through an integrated framework approach?

To answer these questions, this study utilized the methods shown in Figure 1 below:

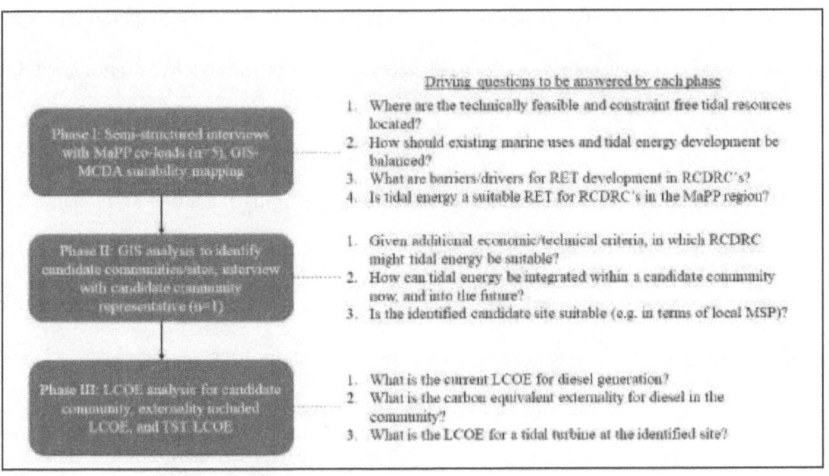

Figure 1: Study methods overview and driving questions for each research Phase.

Integrated assessments and holistic approaches to analyze capability and suitability have been called for in numerous studies and this book exemplifies this paradigm shift, from distinct areas of research, into a cohesive and encompassing framework to assess suitability (Dalton et al., 2015; Jenkins et al., 2018; Jenniches, 2018; Kerr et al., 2014a; Segura et al., 2017b; Segura et al., 2018; Uihlein & Magagna, 2016). In doing so, this research will highlight the benefits of examining such issues through a geography-based lens; combining knowledge and methods from often seemingly disparate disciplines, to investigate potential solutions at numerous scales. As Sheppard so eloquently expressed:

"Geography's intellectual range, from the humanities to the natural sciences, exceeds any attempt to suborn it into a categorical structure of knowledge production. Of course, this very disciplinary structure repeatedly has been challenged by initiatives fostering interdisciplinarity, through cross-cutting curricula, programs, centers, institutes, and clusters. Such initiatives make geography attractive as "the interdisciplinary discipline" whose members (we argue) are uniquely suited to such projects." (2015, p. 1114).

This quote and the methods outlined in this study exemplify the utility of a geography-based approach to complex problems involving multiple considerations at a range of scales. In applying these methods, this research will not only identify opportunities for tidal development near BCs RCDRC, but it will also create an integrated framework for assessing tidal development that can be altered and built upon for global use.

Chapter II: Literature Review

2.1   Tidal energy

2.1.1   Resource and technology overview

Tidal stream turbines capture the energy generated from the gravitational and centrifugal forces of the earth, moon, and sun (Brosche & Schuh, 1998; Segura et al., 2017b). Devices harness the kinetic energy of tidal currents, which are enhanced as tidal waves ebb and flow through constrained passages and inlets (Bedard et al., 2010; Nash & Phoenix, 2017; O'Rourke et al., 2009).

Much of the research and development of tidal stream devices has occurred at test centers such as the European Marine Energy Center (EMEC) in Scotland and Canada's Fundy Ocean Research Center for Energy (FORCE) (Borthwick, 2016; Marine Renewables Canada 2018). Currently, an estimated 100 tidal energy companies are operational worldwide, with many of the most prominent utilizing these test centers (Haslett et al., 2018). In 2018 there was over 20 MW of demonstration and commercial pilot projects deployed globally (Lamy & Azevedo, 2018). By 2022, 1,600 MW of commercial phase projects are slated for operation (Lamy & Azevedo, 2018). The technology readiness level (TRL) for tidal stream turbines (TSTs) is as high as 8 for horizontal axis turbines, with an industry range between 6-8 (Magagna, 2019). Larger scale projects, such as Scotland's Meygen Phase 1A array deployed in 2018, are on the cusp of completeting the TRL path (Magagna, 2019). A multitude of smaller scale devices are also being developed (Marine Renewables Canada, 2018).

There are approximately 100 different TST device concepts falling into five primary device type categories (see Figure 2): horizontal axis devices with parallel axis to the flow, vertical axis devices, horizontal axis devices with perpendicular axis to the flow, oscillating hydrofoil, and other (OES, 2020; Segura et al., 2017b). The trajectory of the tidal stream industry appears to be converging on horizontal axis turbines as the standard device type (OES, 2020). Depth based classification falls into two categories, first generation devices (sea floor moored devices at depths up to 40 meters) and second-generation devices (either floating or submerged to a specific depth) (Segura et al., 2017b). Devices can be kept in place via a monopile, drilled pilot foundations/anchors, or by their own gravity (Segura et al., 2017b).

Figure 2: Typical tidal stream device types and examples: A) Monopile mounted horizontal axis device with perpendicular axis to flow [http://www.siemens.co.uk/en/news_press/pictures/seagen-tidal-current-turbine.htm]; B) Horizontal axis turbine with perpendicular axis to flow (either gravity or piloted) [http://www.orpc.co/our-solutions/turbine-generator-unit]; C) Piloted oscillating hydrofoil [http://www.alternative-energy-tutorials.com/tidal-energy/tidal-power.html]; D) Floating vertical axis turbine with parallel axis to flow [http://www.alternative-energy-tutorials.com/tidal-energy/tidal-power.html] and; E) Anchored tidal kite [http://kis-orca.eu/renewable-energy/wave-tidal-devices/tidal-device-principles].

2.1.2 Advantages

The most apparent obstacle impeding the transition to RETs is the variability of weather-dependent sources of energy and their intermittency, coupled with the difficulties associated with supplying such power to the grid or storing it for future use (Kuang et al., 2016). Tidal energy is predictable with high accuracy over both short and long timeframes, especially when compared to stochastic weather dependent RETs (e.g. solar, wind, and run-of-river) (Johnson et al., 2018; Sangiuliano, 2017a; Sangiuliano, 2017b; Uihlein & Magagna, 2016; Villate et al., 2020). TSTs also benefit from being independent of factors such as rain, fog, or cloud cover (Segura et al., 2017b).

Devices are characterized by high capacity factors, that is, the ratio of actual electricity generated to the maximum amount of electicity that could be generated over a period of time based on the nameplate capacity of a device (Sangiuliano, 2017b; Stothers & Klaptocz, 2016). Although capacity factor can vary by device and site resource characteristics, the technologies on average can be viewed as having promising values in the range of 30-54% (Stothers & Klaptocz, 2016). Results from the first two years of Meygen Phase 1A's operation (4 x 1.5MW horizontal axis turbines) demonstrated capcity factors of 40% and 34% at 100% and 95% availability respectively (average availability of 95%, ranging from 90% in the winter and 98% in the summer) (Black & Veatch, 2020). Compartively, the aggregate 2018 capacity factors for utility scale photovoltaic solar and wind in the United States were 26.1% and 37.4% respectively (EIA, 2019). These characteristics make TST energy a reliable and attractive baseload provider for small scale grids, with the capability to smooth out the cumulative power supply from renewables (Johnson et al., 2018).

Existing studies have shown that tidal stream is favourably viewed in terms of social suitability, including both social benefits and public willingness-to-pay for tidal energy research (Lamy & Azevedo, 2018; Polis et al., 2017). This may be attributed to a range of factors such as submerged devices having no visual impact and no audible sound from devices at or above the surface (see following section for sub-surface sound) (Zaunbrecher et al., 2018). However, opinions regarding social suitability are likely substantially disparate over small and large spatial scales, highlighting the need for stakeholder engagement within suitability studies.

TSTs have thus far demonstrated minimal environmental impacts, as shown through operational monitoring, field studies, and modeling (Bonar et al., 2015; Copping & Hemery, 2020; Hastie et al., 2018; Jenkins et al., 2018; Nash & Phoenix, 2017; O'Carroll et al., 2017; Pine et al., 2019; Ponsoni et al., 2018; Sangiuliano, 2017b; Segura et al., 2018; Uihlein & Magagna, 2016). These impacts range from physical presence (blade strikes on marine fauna and habitat alteration from devices and associated moorings), changes to biophysical properties from energy extraction, underwater noise, chemical pollution, and electromagnetic magnetic fields (Bonar et al., 2015; Copping & Hemery, 2020; Hastie et al., 2018; Jenkins et al., 2018; Nash & Phoenix, 2017; O'Carroll et al., 2017; Ponsoni et al., 2018; Sangiuliano, 2017b; Segura et al., 2018; Uihlein & Magagna, 2016). All these potential stressors have been demonstrated to be low risk for small scale and single device deployments, with reccomendations to move forwad with risk retirement in terms of the regulatory scope and costs required to prove otherwise (Copping & Hemery, 2020; Copping et al., 2020). Many of these stressors are alleviated due to the slow rotational speeds of TSTs (5-70 rpm) (Copping, 2018). The deployment of TSTs is even associated with ecosystem benefits ranging from the provision of habitat to the creation of de facto marine reserves and reductions in flow velocities allowing species to save energy while foraging (Bonar et al., 2015; Haslett et al., 2018; Uihlein & Magagna, 2016).

2.1.3   Disadvantages

Despite the many technical, social, and environmental benefits of tidal energy at small scales, there are disadvantages associated with the industry. Some of these issues are dependent upon scale, while others are the result the state of the industry and characteristics of the marine environment.

While the environmental impact of small scale TSTs and arrays have proven to be minor, questions remain regarding the cumulative risk and magnitude of impacts associated with large scale arrays. Of primary concern is the impacts of cumulative noise inputs, while issues regarding the effects on species ability to navigate, alterations to energy flux, predator-prey interactions, and more remain (Bonar et al., 2015; Pine et al., 2019; van Hees, 2019).

TSTs must also contend with the physical characteristics of the marine environment. Devices are subject to significant forces from both the speed of flows, substantial pressure (seawater is 800 times denser than air), and corrosion due to saltwater (Borthwick, 2016). Furthermore, biofouling is a ceaseless challenge, with potential negative impacts on device function and operation (Borthwick, 2016). Finally, exposure to a range of marine factors such as waves, storms, tidal currents, fog and more may impact the ability to access devices for scheduled maintenance or emergency repairs.

Issues regarding the policies and regulatory structures in place to govern the industry is also an area of concern (Andersson et al., 2017; Richardson, 2018). This creates difficulties in standardizing the industry, along with unforeseen costs and obstacles to potential deployments and development. Many countries around the world lack specific legal and regulatory frameworks for tidal energy, including Canada, compounded by the fact that regulatory bodies are often risk adverse, especially towards novel technologies (Richardson, 2018; Uihlein & Magagna, 2016). The result of such uncertainty and lack of regulatory guidance can result in permitting processes not fit for purpose, creating barriers to the development of the industry (Andersson et al., 2017; Wright 2015).

While TSTs are certainly technically feasible, the industry is still working towards becoming economically and commercially viable (Johnson et al., 2018). Costs and risks are greatest in this phase between device design/testing and commercial viability, known colloquially as the 'Commercial Valley of Death' (Villate et al., 2020).

2.2   Diesel generated electricity

The remoteness of many Canadian communities made the costs of connecting to the electric grid unfeasible and instead the Canadian government-initiated programs to electrify remote communities with hydroelectric and diesel plants in the 1960s and 70s (Karanasios & Parker, 2018). The federal government was typically responsible for the capital costs of generators, while provincial governments/utilities were responsible for the operation and maintenance of community power plants while providing electricity at a reasonable price (Karanasios & Parker, 2018).

2.2.1 Technology overview

Diesel engines were first patented in 1892 by German engineer Rudolf Diesel (EIA, 2020). Since then, they have been applied to a multitude of uses from land, sea, and air transportation to heat and electricity generation. Diesel generators take refined crude oil and convert the chemical energy trapped within hydrocarbons into mechanical energy via combustion. This mechanical energy is then used to spin an alternator, creating AC electrical current.

2.2.2 Advantages

One of the greatest advantages of diesel generators, and other fossil fuel-based electricity, is their ability to provide power on demand. This makes electricity production easily plannable. Evidence of this dispatchability is evident in the use of diesel generators as backup power sources for critical infrastructure (e.g. hospitals, data centers) in case of grid outages or natural disasters. Diesel generators range in size from a few kW to MW, which allows them to be deployed in remote applications for a single building to entire communities. Furthermore, multiple generators can be integrated into grids to meet increased demand. Diesel generators are cheaper than other forms of fossil fuel based electrical generation, such as gas-powered turbines, while also being far cheaper than connecting remote communities to the grid (Karanasios & Parker, 2018). With many communities reliant on diesel power since the late 20th century, their familiarity with diesel generators in terms of operational and maintenance requirements along with an overall understanding of the technology make them appealing to many communities that often find themselves isolated and without support if systems fail.

2.2.3 Disadvantages

While diesel generators are a dispatchable source of electricity which are apt to be deployed in remote communities, their use is associated with several disadvantages. With many generators already operating at full capacity, the likelihood of enhancing living standards is severely constrained, as proposed community projects or upgrades may have substantial electricity requirements (Arriaga et al., 2016; Karanasios & Parker, 2018; Kennedy, 2018; Price Waterhouse Cooper, 2015). This in turn may further exacerbates social issues in what are already often disproportionately disadvantaged communities (Arriaga et al., 2016; Statistics Canada, 2016). Existing generators are often run counterintuitively to device longevity (e.g. not run at their optimal rated output) to meet fluctuating daily demand, resulting in brownouts and the risk of generator failures and blackouts (Kennedy, 2018). Furthermore, with diesel fuel being shipped to the remote communities from tens to hundreds of kilometers away, energy security and community independence are substantially decreased (Arriaga et al., 2013; Kennedy, 2018).

The principle environmental issue with diesel electricity generation are GHG emissions. Remote off grid communities have nearly three times Canada's per capita average emissions for electricity and heat generation, further compounded by the transportation of fuel to communities using vehicles that themselves consume diesel (e.g. tug and barges) (ECCC, 2016; Kennedy, 2018; NRCan, 2011). Diesel fuel also presents a risk to the environment, with the potential for acute spills during transport and or chronic leaks while being stored (NRCan, 2011; Transportation Safety Board of Canada, 2018). Fuel tank leaks can directly impact communities through the contamination of soil and groundwater, while marine transportation in BC threatens some of the most productive and sensitive ecological habitats on earth of which many communities are reliant upon for sustinence living and economic opportunties (MaPP, 2016;

NRCan, 2011; PNCIMA, 2017). Such fears have been realized in the MaPP region already, for example the sinking of the Nathan E Stewart tugboat in the Heiltsuk Nation's traditional territories (Heiltsuk Tribal Council, 2017). The tug released 111,000 L of diesel into the marine environment resulting in fishing closures, impacts to cultural activities, substantial clean up costs, marine fauna and flora mortality, and long term damage (Heiltsuk Tribal Council, 2017). Luckily the 11.7 million liter capacity fuel barge was empty, greatly reducing the potential magnitude of the event (Heiltsuk Tribal Council, 2017).

Economic issues with diesel electricity generation abound ranging from price volatility (see Figure 3 below); to high operation, maintenance, and transportation costs (Arriaga et al., 2013; McFarlan, 2018; NRCan, 2011). The price of producing diesel generated electricity in remote Canadian communities in 2011 ranged from $0.51 to $2.82/kWh, with RCDRC such as Hartley Bay having generation costs around $0.74/kWh (Arriaga et al., 2014, NRCan, 2011; NRCan, 2013). Comparatively the average price of electricity in the rest of Canada was between $0.06 and $0.19/kWh (values expressed in 2020 CAD based on ten-year average currency conversions, adjusted for 1.35% inflation per year) (Bank of Canada, 2020; NRCan, 2011).

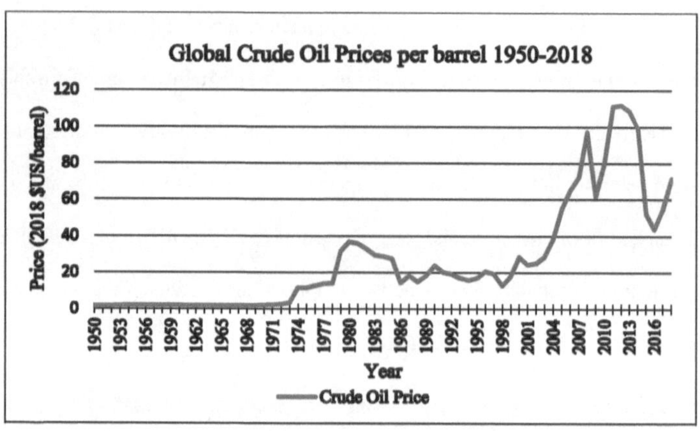

Figure 3: Global crude oil prices over the past 68 years. Expressed as 2018 US dollars per barrel. Data taken from: OurWorldInData.org/fossil-fuels and sourced from BP 2018 Statistical Review of World Energy.

2.3   Stakeholder engagement

A holistic approach towards renewable energy development research necessitates the involvement and input of local stakeholders (Arnstein, 1969; Rowe & Frewer, 2000). The development of tidal energy reflects this, with increasing calls within the academic community to incorporate human dimensions into evaluations (Jenkins et al., 2018). So far, the human dimensions of tidal energy development have received far less attention from the scientific community than resource assessments, evaluation of environmental impacts, and device design (Kerr et al., 2014b; Ruano-Chamorro et al., 2018). Arising from this attention deficit is a lack of public understanding, resulting in the inability for stakeholders to create informed opinions regarding acceptability and to formulate an understanding of the interplay between tidal energy and their way of life (Dalton et al., 2015; OES, 2020; Rowe & Frewer, 2000; Ruano-Chamorro et al., 2018).

Public perceptions about potential risks associated with renewable energy is dependent upon their understanding of the technology and the social values within their community (Dalton et al., 2015). While factors such as costs are key considerations for understanding the feasibility of a potential energy project; public understanding, local support and acceptability remain crucial to ensuring the project is successfully developed and that it satisfies community needs (Dalton et al., 2015; Frazão Santos et al., 2018; Kerr et al., 2014b; Ruano-Chamorro et al., 2018).

The inclusion of human dimensions into tidal energy assessments necessitates stakeholder engagement, which is widely regarded as a key component of contemporary spatial planning (Arnstein, 1969; Kerr et al., 2014b; Rowe & Frewer, 2000; Quero García et al., 2019). Stakeholder engagement has developed into a rich area of academic discourse ever since Arnstein's ladder of engagement paper highlighted the need to provide stakeholders with the power to have influence over planning and development decisions (Arnstein, 1969; Irwin, 2006; Rowe & Frewer, 2000). Despite a half century of discourse, there is still no universal way of developing and enacting engagement along with no proven method to do so (Cuppen et al., 2016; Irwin, 2006; Rowe & Frewer, 2005; Webler & Tuler, 2002; Whitman et al., 2015).

What has become apparent, time and time again, is the need to develop stakeholder engagement methods tailored to specific projects (Dyer et al., 2014; Pomeroy & Douvere, 2008; Ritchie & Ellis, 2010; Rowe & Frewer, 2005). While taking and adapting methods from cases of successful engagement is a promising start, researchers must be cognisant of stakeholder feedback and incorporate it into ongoing and iterative discussions (Dyer et al., 2014; Pomeroy & Douvere, 2008; Ritchie & Ellis, 2010; Rowe & Frewer, 2005). Studies have highlighted the need to balance relationships between researchers and stakeholders, early and continuous engagement, a process built on trust, and most of all to create engagement anchored in co-learning (Arnstein,

1969; Frazão Santos et al., 2018; Reed, 2008; Ritchie & Ellis, 2010; Rowe & Frewer, 2005). Researchers must also be cautious of temporal and financial holds on their methods and acknowledge that it may not always be feasible or necessary to include every stakeholder in a given study (Rowe & Frewer, 2000). Successful engagement may also require the use of proxies for larger groups of stakeholders, for instance, representative bodies in planning initiatives or community leaders (Rowe & Frewer, 2000).

Engaging and collaborating with remote communities in BC, and Canada more broadly, requires flexibility and adaptability. Despite many communities sharing similar characteristics, challenges facing communities and their capacity to solve them vary in magnitude (Knowles, 2016). The optimal energy solution for each remote community will likely be community specific, dependant on limitations such as funding/financing availability, climate, population size, local resources, geography, human resource capacity, and more (Knowles, 2016). This highlights one of the other benefits of early engagement, the identification of community strengths, weaknesses, and specific needs.

TST project engagement not only enhances the likelihood of a project being accepted, as shown in the development of onshore renewables, it also facilitates the collection and incorporation of social data into broader marine planning (Frazão Santos et al., 2018; Kerr et al., 2014b). The inclusion of stakeholder values and perceptions in tidal energy developments will provide benefits to the local community beyond energy provision such as job creation, enhanced independence/self sufficiency, and the protection of place-based values (De Groot & Bailey, 2016; Kerr et al., 2014b; Ruano-Chamorro et al., 2018; Segura et al., 2017b).

2.4   Analyzing the spatial feasibility and suitability of tidal energy

2.4.1   Marine Spatial Planning

MSP defined as "…a public process of analyzing and allocating the spatial and temporal distribution of human activities in marine areas to achieve ecological, economic, and social objectives usually specified through a political process," is a promising method for delineating and balancing uses of the marine space in a holistic manner (Ehler & Douvere, 2009, p. 18). The development and implementation of MSP has been initiated in direct response to increasing anthropogenic pressures on marine resources and the expansion of uses within the marine space over the past half century (Kerr et al., 2015; Tawil et al., 2018; Wright, 2015). MSP has been developed in approximately 70 countries thus far with overarching goals to balance human uses, environmental protection, and the preservation of social values in the marine space (De Groot & Bailey, 2016; Ehler & Douvere, 2009; Frazão Santos et al., 2018; Jenkins & Dreyer, 2018; Kerr et al., 2015; Quero García et al., 2019).

Although MSP views uses within the marine space and the balancing of them largely through a cohesive lens, it has shown great promise in the EU, especially in Scotland, to facilitate and enable the development of MRE (Quero García et al., 2019; Richardson, 2018). Experiences with offshore wind and onshore renewables suggests that MRE will likely encounter resistance from marine activities, especially when operating near coastal communities where anthropogenic activities tend to congregate (Kerr et al., 2014; Kerr et al., 2015; Segura et al., 2018). Furthermore, MRE development necessitates the 'ownership' or proprietary use of areas of the marine space resulting in the exclusion of others (Kerr et al., 2015). Consequently, planning for TST development requires an understanding of the complexity of biophysical and human dimensions of ocean uses in order to integrate the multi-level management objectives encapsulated within MSP frameworks (Frazão Santos et al., 2018; Haslett et al., 2018; Segura et al., 2018).

Despite minimal implementation of legislated MSP frameworks enabling MRE in Canada, apart from Nova Scotia, and with large scale planning efforts in BC lacking direct federal involvement/support at this time, MSP processes and tools can still be used to develop and enhance marine management (Frazão Santos et al., 2018; MaPP, 2016; Marine Renewables Canada, 2018; Richardson, 2018).

2.4.2 Geographic Information System Multi-Criteria Decision Analysis

Decision making for renewable energy siting is a complex and often convoluted task owing to the multitude of social, economic, technical, and biophysical factors which in turn yield suitable development sites. For instance, while in the simplest terms a suitable site requires sufficient tidal resources, a myriad of additional factors such as: environmental, economic, and social uses; site proximity to demand and large ports, depth, and more contribute to a more encompassing assessment of suitability (Defne et al., 2011; Galparsoro et al., 2012; Thomas et al., 2019).

The use of GIS MCDA provides decision makers with a systematic operational evaluation and decision support tool that excels at tackling complex problems with high levels of uncertainty, diverging objectives, multiple interests, and varied information/data forms (Giamalaki & Tsoutsos, 2019a; Maslov et al., 2014; San Cristóbal, 2011; Vasileiou et al., 2017; Wang et al., 2009). GIS-MCDA has the potential to minimize project costs, reduce conflicts with other uses, minimize environmental impacts, and avoid stakeholder opposition (Defne et al.,

2011; De Groot & Bailey, 2016; Jenkins et al., 2018; Quero García et al., 2019; Uihlein & Magagna, 2016; Vasileiou et al., 2017; Wang et al., 2009).

Geographic Information Systems

GIS has been commonplace over the past twenty years to assess constraints and select suitable sites for energy projects, thanks to its ability to visually represent spatial data and the suite of processing tools available (Cradden et al., 2016; Defne et al., 2011; Janssen et al., 2014; Kim et al., 2018). Spatial mapping and analysis using GIS is also a commonly used tool within broader MSP efforts that can effectively, both from a cost and temporal perspective, represent the marine space and delineate uses within it (Frazão Santos et al., 2018; Marine Renewables Canada, 2018; Richardson, 2018). However, the effective utilization of GIS is dependent on the availability of data sets to represent each criterion, with immense challenges in terms of temporal and financial cost to produce necessary data sets if they are not available.

With TST costs being highly dependent upon site location, mapping exercises will aid in the financial appraisal of newly considered developments (Segura et al., 2017b). However, merely representing indicators of site suitability for disparate criteria does not simplify decision making, as these layers still need a foundation for which to combine them and analyze trade-offs, which is where MCDA comes into play.

Multi-Criteria Decision Analysis in renewable energy technologies siting

MCDA encompasses a range of decision-making tools that enable the investigation of relationships between multiple criteria to examine trade-offs and achieve a predefined optimization objective or suitability assessment (San Cristóbal, 2011). MCDA approaches are suitable for energy system evaluations as they involve multiple decision makers, contradictory criteria, along with being subject to long time frames, varied sources of uncertainty, and capital-intensive investments making initial siting crucial (Giamalaki & Tsoutsos, 2019).

MCDA methods are primarily categorized as multi-objective decision making (MODM) and multi-attribute decision making (MADM) methods (Höfer et al., 2016; Kurka & Blackwood, 2013). The main differentiation between the two is the number of possible alternatives evaluated (e.g. either being continuous or discrete) (Höfer et al., 2016; Kurka & Blackwood, 2013; Zanakis et al., 1998). MODM represents a design approach, in which the most optimal solution (e.g. suitable site) is defined by parameters (e.g. site suitability criteria) and determined within a infinite set of solutions (Höfer et al., 2016; Kurka & Blackwood, 2013; Zanakis et al., 1998). Conversely, MADM methods embody a ranking approach, in which the best solution among a constraint screened finite number of alternatives is chosen based on ranking according to decision making criteria (Höfer et al., 2016; Kurka & Blackwood, 2013; Zanakis et al., 1998). Of the two, MADM based methods are predominantly utilized in RET siting (Kurka & Blackwood, 2013).

MADM techniques utilized for RET site assessments include, but are not limited to: the Analytic Hierarchy Process (AHP), the Elimination and Choice Translating Reality (ELECTRE) method, the Technique for Order Preference by Similarity to Ideal Solutions, the Ordered Weighted Averaging technique, fuzzy MCDA methods, and hybrid MCDA methods (Giamalaki & Tsoutsos, 2019). Selection of an appropriate MCDA technique necessitates evaluating the method's ability to deal with uncertainty, user friendliness and flexibility, transparency of the method (i.e. is it likely to improve stakeholder comprehension) and its ability to include multiple stakeholders (Giamalaki & Tsoutsos, 2019; Kurka & Blackwood, 2013).

Within existing literature, the AHP represents one of, if not the most commonly used MCDA technique for RET site assessments (Giamalaki & Tsoutsos, 2019; Höfer et al., 2016; Kurka & Blackwood, 2013; Mahdy & Bahaj, 2018; Stefanakou et al., 2019; Vasileiou et al., 2017; Wang et al., 2019). The AHPs ability to allow for the combination of different evaluation criteria (both qualitative and quantitative), incorporate stakeholder/expert weighting of criteria, provision of a logical and mathematical justification to decision making, and simplification of the decision making process all attribute to its prevalence (Ali et al., 2018; Giamalaki & Tsoutsos, 2019; Kim et al., 2018; Stefanakou et al., 2019). A more in-depth overview of the AHP methods can be found in section 4.1.1.

2.4.3   GIS suitability mapping and GIS-MCDA mapping for tidal energy

GIS-MCDA involves the representation of identified data inputs as layers within a GIS software. These layers are usually normalised to allow for comparability and analysis of disparate data sets, often on a scale of 0-100 (Ang et al., 2016; Cradden et al., 2016; Davies et al., 2014; Defne et al., 2011; Van Cleeve et al., 2013). These layers are then weighted, that is, given numerical representations of importance, and combined using GIS algebraic functions in order to generate a spatial representation of relative suitability (Ang et al., 2016; Cradden et al., 2016; Davies et al., 2014; Defne et al., 2011; Van Cleeve et al., 2013).

Existing studies

A range of tidal GIS and MCDA studies exist, such as Defne et al., (2011) which assessed tidal stream potential in Georgia USA using GIS based multi-criteria assessment. Davies et al., (2012) undertook a study identifying areas for tidal stream energy development in Scottish waters across technical, industrial, environmental, and socio-cultural themes. Van Cleeve et al.,

(2013) examined tidal energy suitability with a range of technical and economic considerations (not including legal and regulatory constraints). Janssesn et al., (2014) also examined potential sites for tidal energy in Scotland, using stakeholder decision support tools and spatial multicriteria analysis. Maslov et al., (2014) used the ELECTRE III MCDA approach in combination with GIS to create a model and applied it to identify tidal farms in the North West of France based on social acceptance, along with ranking them and estimating cost and energy production. Cradden et al., (2016) examined site suitability for offshore energy platforms more generally while Ang et al., (2016) created a web-based GIS MCDA tool for examining tidal current energy development in the Philippines. Finally, Vazquez & Iglesias (2016a) developed a MATLAB geospatial analysis for the Bristol channel to identify locations for tidal energy and calculate LCOE values while considering technical (resource), economic (shipping), and functional constraints (conservation areas, submarine cables, department of defense areas).

2.5   Economic feasibility

2.5.1   Tidal energy costs

The current understanding, or lack thereof, regarding TST costs impedes the development of the industry as it dissuades private investment, perpetrating an industry heavily reliant on public support (Polis et al., 2017; Sangiuliano, 2017a; Segura et al., 2017b). The infancy of the industry results in tidal being viewed as a risky investment due to its high up-front capital costs, exogenous and endogenous uncertainties, along with the financial demands of iterative technological and processes development (Johson et al., 2018; MacDougall, 2017; Sangiuliano, 2017a; Sangiuliano, 2017b; Segura et al., 2018). It is therefore necessary to build upon existing knowledge and develop studies which include assessments of TST economic viability to attract investments (OES, 2020; Segura et al., 2017a).

2.5.2   Levelized Cost of Energy

LCOE calculations are one of the primary instruments for evaluating the profitability of energy production technologies (Nissen & Harfst, 2019; Segura et al., 2017a). This can be attributed to their ability to compare dissimilar energy production means and nameplate capacities (Nissen & Harfst, 2019; Segura et al., 2017a). LCOE can be defined as the cost per unit energy (usually expressed in \$/kWh or \$/MWh) generated over the lifetime of a project, that is, life-cycle costs divided by lifetime energy production (Bruck et al., 2018; Dalton et al., 2015; Nissen & Harfst, 2019; Segura et al., 2017a; Vasquez & Iglesias, 2016).

Equation 2.1 yields LCOE:

$$LCOE = \frac{Ccapex + \sum_{t=1}^{n} Copex_t \times (1+r)^{-t}}{\sum_{t=1}^{n} E_t \times (1+r)^{-t}} \qquad 2.1$$

Where C_{CAPEX} denotes capital expenditures, C_{OPEXt} is the cost of operational expenditures in year t, E_t is the electricity generation in year t, r is the discount rate, and n is the lifetime of the system (Dalton et al., 2015; Nissen & Harfst, 2019; Segura et al., 2017a; Vasquez & Iglesias, 2016). CAPEX is characterized by benefits that extend beyond one year, mostly accounting for the general costs of an energy system such as the device itself, mooring systems, energy transportation systems, and so forth. (Dalton et al., 2015; Segura et al., 2017). CAPEX is the primary determinant of TST LCOE, accounting for 70% of total LCOE (Vazquez & Iglesias, 2016b). Therefore, CAPEX reductions achieved through device standardization and improved installation methods driven by industry learning will have substantial impacts on future costs

(Segura et al., 2019; Vazquez & Iglesias, 2016b). OPEX represents yearly expenses such as administrative costs, operation and maintenance costs (both scheduled and unscheduled), taxes, and more (Dalton et al., 2015; Segura et al., 2017). E_t is primarily dependent upon nameplate capacity, device capacity factor, and resource potential (Dalton et al., 2015; Segura et al., 2017a). Discount rates are based on two notions: the first being the time value of money (that is money is more valuable today than it is in the future) and the second uncertainty risk (that is cash in the future is associated with risks dependant on the uncertainty of the investment). It essentially represents the rate of return an investor would need to receive to justify the investment, with riskier investments requiring a higher discount rate and vice versa.

Recent TST LCOE assessments from device deployments range from $490-547/MWh (average of $526/MWh for the 10 MW deployed in the UK) (SETIS, 2019; Smart & Noonan, 2018; Villate et al., 2020). Comparatively, the LCOE of offshore wind is currently around $215/MWh (OES, 2015). Despite the significant gap, costs continue to decrease within the TST industry, with the European Commission's Joint Research Centre finding a 40% reduction in tidal streams LCOE over the past three years, reducing costs from an average of $724 ($505-945) in 2015 (OES, 2015; Villate et al., 2020). Future reductions look promising with the European Commission setting cost targets of $216/MWh for 2025 and $144/MWh by 2030 (all values above have been adjusted for inflation and converted to 2020 CAD) (OES, 2019).

2.5.3 Externalities and willingness to pay

While traditional economic assessments provide an understanding of the private costs/benefits of energy production, they often fail to incorporate external costs and benefits over a project's lifetime (Eidelwein et al., 2018; Polis et al., 2017; Sangiuliano, 2017a). These environmental and societal impacts not accounted for by either private or public parties are known as externalities. Externalities can generate benefits (positive externalities) or costs (negative externalities) (Eidelwein et al., 2018; Rezai et al., 2012). Externalities associated with energy production are primarily negative and represent costs to society or the environment that do not impact the economic outcomes of their generating agent (Eidelwein et al., 2018; Rezai et al., 2012). Common examples of negative externalities associated with power production include CO_2 and other GHG emissions, health impacts, visual amenity impacts, and land use changes. Failure to consider externalities can lead to an incomplete understanding of energy systems, impeding holistic management and planning.

Externalities are predominantly expressed in monetary terms (sometimes referred to as environmental accounting), as this format is more accessible to financial institutions and policy makers (Eidelwein et al., 2018; Lohmann, 2009; Streimikiene et al., 2019). Many methodological challenges are associated with quantifying these costs and benefits, especially given the long-time frames they are expressed over and uncertainties associated with the magnitude of future climate change and impacts (Allen et al., 2018; Jenniches, 2018; Sovacool et al., 2020; Streimikiene et al., 2019). Valuating environmental externalities can involve a multitude of sources including market values, scientific studies, modeling, and more (Eidelwein et al., 2018).

Chapter III: Study area overview

3.1   Study area

3.1.1   Existing MSP

This study will utilize and build upon the foundations of existing MSP in BC, being

conducted within the spatial bounds of the Pacific North Coast Integrated Management Area

(PNCIMA) and the Marine Plan Partnership for the North Pacific Coast (MaPP). The

geographical boundaries of these plans encompass nearly two thirds of BCs coastline (see Figure

4) (MaPP, 2016; PNCIMA, 2017).

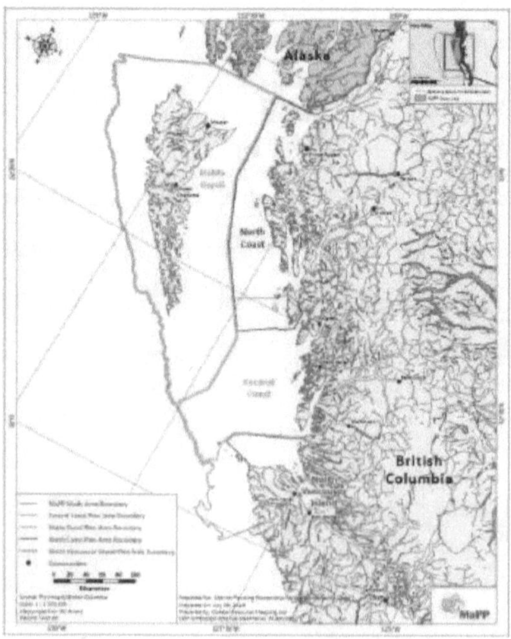

Figure 4: Marine Planning Partnership for the North Pacific Coast plan area (Coastal Resource Mapping Ltd, 2015).

PNCIMA was the first MSP initiative in BC, initiated in 2009 and involving four levels of government: federal, First Nations, provincial, and local (PNCIMA, 2017). Federal support for the plan was withdrawn for a time in 2012, and in response, MaPP was created. MaPP is a co-led process between the provincial government (represented by the Ministry of Forests, Lands, Natural Resource Operations and Rural Development, FLNRO) and eighteen First Nations (italicised in Table 1) to foster opportunities for economic development, support the well-being of coastal communities, and protect the marine environment (MaPP, 2016).

The MaPP regional and sub-regional plans will be taken as the principal MSP reference in this study as they represent one of the more recent and encompassing MSP initiatives in BC and included substantial First Nations involvement in their creation. Four different sub-regional plans provide the spatial context for the regional plan (MaPP, 2016). Each of the sub-region plans were individually created by a combination of local First Nations and representative conglomerates, the provincial government, and sub-regional planning committees (see Table 1).

Zoning further delineates each sub region into three primary categories: General Management Zones (GMZ) in which multiple uses and activities are allowed; Special Management Zones (SMZ) in which specific management emphasis is put on uses such as recreation and tourism, cultural, cultural/economic, community, aquaculture, and alternative energy; and Protection Management Zones (PMZ) which prioritize conservation objectives and compatible uses (MaPP, 2016).

Table 1: Sub-regional plan partners and First Nations involved in the MaPP.

Sub-region	Partners (bold represents First Nations conglomerates)	Individual First Nations represented
North Vancouver Island	BC Government	
	Nanwakolas Council	*Da'naxda'xw Awaetlatla*
		Gwa'sala'Nakwaxda'xw
		K'omoks
		Kwiakah
		Mamalilikulla-Qwe'Qwa'Sot'Em
		Tlowitsis
		Wei Wai Kum
	NVI Marine Plan Advisory Committee	
Central Coast	BC Government	
	CC Marine Adivsory Committee	
	Heiltsuk	
	Kitasoo/Xai'Xais	
	Nuxalk	
	Wuikinuxv	
North Coast	BC Government	
	NC Marine Plan Advisory Committee	
	North Coast-Skeena First Nations Stewardship Society	*Gitga'at*
		Gitxaala
		Haisla
		Kitselas
		Kitsumkalum
		Metlakatla
Haida Gwaii	BC Government	
	Council of the Haida Nation	*Haida*
	HG Marine Advisory Committee	
	Haida Marine Working Group	

3.1.2 Environment brief

Much of the study areas coastline is composed of fjords carved into the granitic Coast

Mountains during several periods of glaciation, the last ending 12,000 years ago (Johannessen et

al., 2007). This glacial activity along with millennia of subsequent erosional and depositional geomorphic processes has produced a coastline characterized by numerous offshore islands and archipelagos, rocky shores, sand/gravel beaches, and estuaries (PNCIMA, 2017; Johannessen et al., 2007). The study area is generally dominated by a coastal temperate climate characterized by mild temperatures and high precipitation as a result of its proximity to the Pacific Ocean, prevailing winds, and the orographic effect of the coastal mountains (Johannessen et al., 2007). The primary synoptic scale features influencing year-round weather are the Aleutian Low and the North Pacific High (Johannessen et al., 2007). The winter months are typically dominated by the Aleutian Low, bringing storms and strong southeast to southwest winds. Conversely the North Pacific High dominates in the summer with lighter winds from the northwest and fairer weather.

The study area is home to a diverse and rich array of marine life for at least a portion of their life history thanks to consistent upwelling of nutrient rich water and strong tidal mixing (PNCIMA, 2017). This diversity includes bird populations such as gulls, eagles, cormorants, puffins, storm petrels, black oystercatchers, ancient murrelets, auklets, and many more (Johannessen et al., 2007). Marine mammals also abound ranging from cetaceans such as northern resident, offshore and transient killer whales; Pacific white-sided dolphins, grey whales, harbour porpoises and more; to pinnipeds including sea otters, Stellar sea lions, harbour seals, and Northern fur seals (Johannessen et al., 2007). Many more species occupy the offshore waters of the study area such as transient and offshore killer whales; or pass through the waters on migratory routes such as blue, sei, fin, sperm, and northern right whales (Johannessen et al., 2007). There are also numerous fish and invertebrate species such as, but certainly not limited to: clams, octopus, flounder, hake, herring, lingcod, salmon (coho, pink, chinook, sockeye, chum, and steelhead), pollock, rockfish, prawns, scallops, shrimp, urchins, sea cucumbers, and crabs (MaPP, 2016).

The area has an extensive watershed intertwining marine and terrestrial ecosystems over 177,000km^2 (Johannessen et al., 2007). There are also a multitude of benthic habitats producing fertile nurseries which yield thriving ecosystems. These range from rocky reefs to muddy sediments and nearshore gravel beds, providing habitat for an array of ecological communities often characterized by a single species such as eel grass or kelp (Johannessen et al., 2007; MaPP, 2016). Fifteen Ecologically and Biologically Significant Areas can be found within the region, delineated as such based on their ecosystem rarity, life stage importance to species, vulnerability, productivity, and diversity (Johannessen et al., 2007).

Despite the abundance of marine life in the region, there have been significant concerns regarding dwindling fish populations (especially salmonids) along with shipping noise underwater among others (MaPP, 2016; PNCIMA, 2017). Like the rest of the world, the study region will experience and has been experiencing the effects of climate change. Air temperature is projected to increase 1.8°C by 2050 and 2.7°C by the 2080's driving impacts ranging from ocean acidification, increasing sea temperature, and deoxygenation to declines in sea surface salinity, sea level rise, and more (Whitney & Conger, 2019). The cumulative effects of these changes are expected to alter the marine environment, with a projected 30% reduction in total ecosystem biomass within the food webs of the northeast Pacific (Whitney & Conger, 2019).

3.1.3 Social brief

The study regions population exceeds 100,000, distributed across 14 incorporated, 18 unincorporated (i.e. not governed by a local municipality), and 32 First Nations communities (MaPP, 2016). The four largest urban centres are Campbell River, Prince Rupert, Terrace, and Kitimat; with Campbell River being the largest with a population of 32,000 (MaPP, 2015d).

First Nations peoples are inextricably linked and engrained within the study region having habited it for over 11,000 years (MaPP, 2015c). These Nations have utilized and stewarded both the marine and terrestrial areas of the region since time immemorial and continue to do so today. Their uses include subsistence living, commercial resource exploitation, environmental monitoring/management, along with innumerable social and cultural activities (MaPP, 2016; PNCIMA, 2017). The extent of their traditional territories and populations were substantially decreased due to colonial contact (Kennedy, 2018; MaPP, 2015d). Colonial governance excluded First Nations from participating in the economy of the day while stifling cultural and traditional activities (MaPP, 2015d). As a result, the MaPP regional plans have significant focus on improving socio-economic conditions within First Nations communities along with increasing their participation in the regional economy (MaPP, 2015d; MaPP, 2016).

Though it is paramount to acknowledge, sufficient outlining and explanation of the immediate, cumulative, and continuing effects of colonialism upon these Nations and others in BC and Canada is beyond the scope of this book. Fitzgerald and Kennedy each delve deeper into the effects of colonialism and the disadvantages produced in First Nations communities (Fitzgerald, 2018; Kennedy, 2018).

3.1.4 Economic brief

The natural richness and abundance of the region has enabled the development of a plethora of economic activities within the marine space. Marine economic activities within the region are primarily characterised by resource-and nature-based activities (MaPP, 2016). These include, but are not limited to; aquaculture, commercial fishing, forestry, shipping, recreation, and tourism (Johannessen et al., 2007; MaPP 2015a; MaPP 2015b; MaPP 2015c; MaPP 2015d; MaPP, 2016; PNCIMA, 2017).

All regional plans seek to enable sustainable economic development in order to maintain populations and continue to provide adequate livelihoods for the inhabitants of the region (MaPP, 2016). Such actions necessitate the identification of, and focus upon, more sustainable industries while clearly illustrates a desire to pursue economic development focused on long-term sustainability (MaPP 2015a; MaPP 2015b; MaPP 2015c; MaPP 2015d). Many proposed economic activities have divided opinions both within the study region and BC more broadly, primarily the expansion and development of fossil fuel projects (MaPP, 2015c). MRE is cited in each regional plan as a potential new source of economic development and is supported by the identification MRE SMZs (MaPP 2015a; MaPP 2015b; MaPP 2015c; MaPP 2015d).

Chapter IV: Phase I methods

4.1   Interviews and GIS-MCDA

Phase I of this study focused on developing a GIS-MCDA framework to identify suitable sites for tidal energy development. This was achieved through the determination of appropriate weights for each 'suitability' sub-model via semi-structured interviews with the planning co-leads for each MaPP sub-region and by using the AHP and GIS *Weighted Overlay* mapping. The flowchart overview of methods utilized in Phase I can be found in Figure 5 below. Layers represent datasets used, sub-models represent determinants of suitability, and models refer to final products for examining suitability.

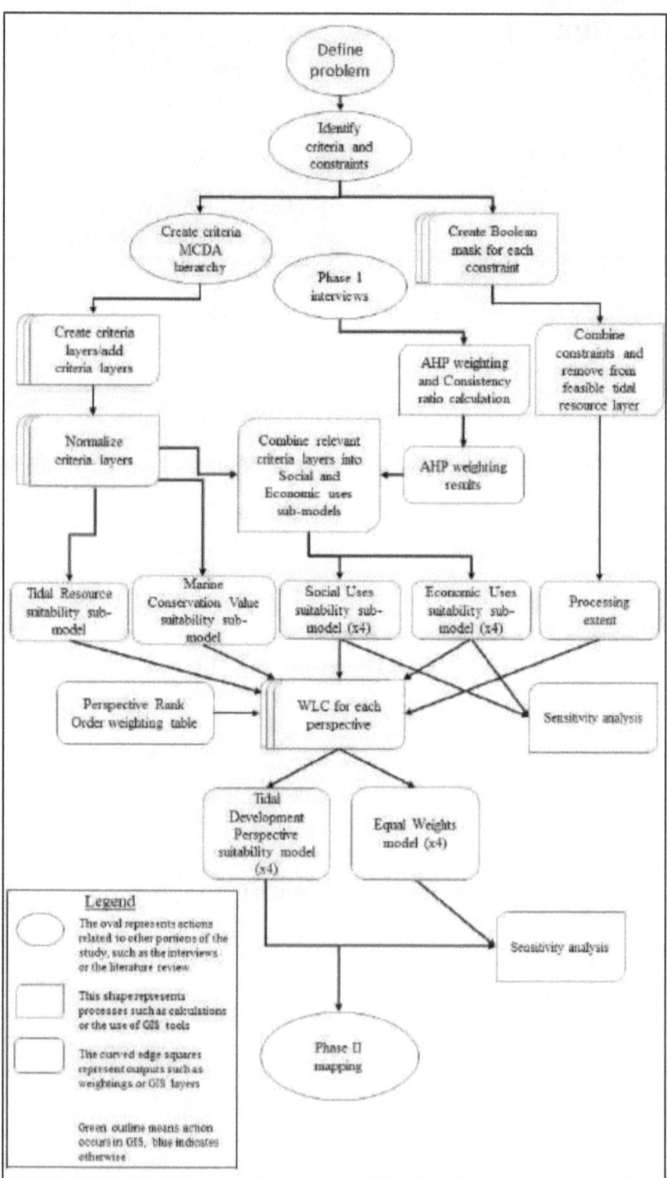

Figure 5: Phase I GIS-MCDA methods flowchart.

The interview questions (see section 5.1) were developed in order to understand how tidal energy is currently viewed within the MaPP region, barriers to development, how tidal energy might be integrated in the future, along with the AHP pair-wise comparison matrices (see Appendix G) to determine criteria weights. Questions were developed using a mixed methods approach designed to yield quantitative and qualitative results for comparability and to elicit responses that elaborated on the answers provided.

The FLNRO and First Nations co-leads, or a representative of the First Nations planning conglomerate for each sub-region were identified and contacted. In total, 5 interviews with co-leads or representatives were undertaken: two FLNRO Co-leads (Haida Gwaii marine plan and the co-lead for both the Central Coast and North Vancouver Island marine plans), the Central Coast Indigenous Resource Alliance Co-lead (Central Coast marine plan), a representative of the Nanwakolas Council (North Vancouver Island marine plan), and the Council of the Haida Nation (CHN) Co-lead (Haida Gwaii marine plan).

The North Coast marine plan FLNRO co-lead was unable to participate and there was not a First Nations Co-lead at the time. To gain some insight into the North Coast sub-region, an interview was undertaken with a representative of the Metlakatla First Nation. While the additional participant provided valuable input and insight, their role and characteristics of the community they represented (grid connected rather than diesel powered) were not consistent with the framework of this study and thus some of their answers (particularly AHP responses) were excluded.

4.1.1   The Analytic Hierarchy Process

There are four primary steps in implementing the AHP, the first being the creation of a hierarchical structure of the siting problem (Giamalaki & Tsoutsos, 2019; Höfer et al., 2016; Mahdy & Bahaj, 2018; Saaty, 1980; Stefanakou et al., 2019; Wang et al., 2009). This entailed first defining the problem and specifying which criteria should be involved in decision making.

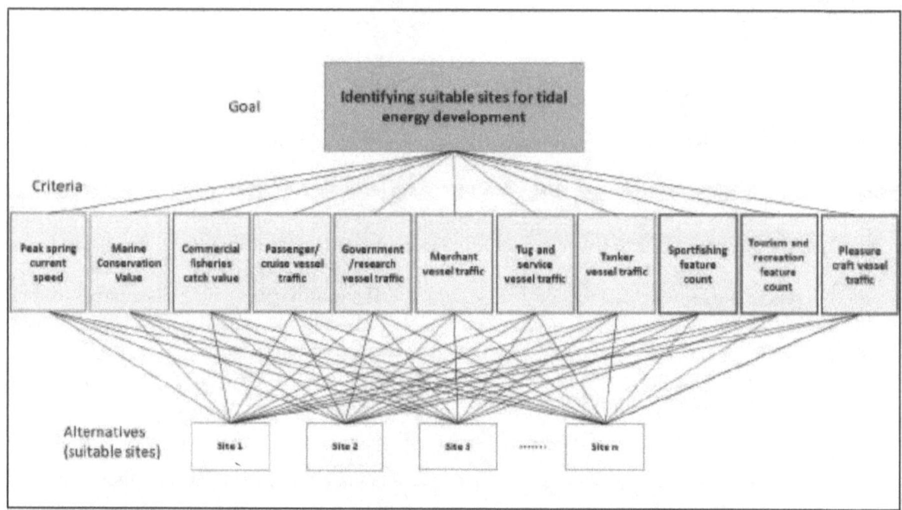

Figure 6: AHP decision hierarchy to identify suitable locations for tidal energy development.

Within this hierarchy (Figure 6) the goal of the problem (e.g. identifying suitable sites for tidal energy development) occupies the top level, the evaluation criteria (e.g. resource feasibility along with economic, social, and environmental considerations) is situated in the middle, and the alternatives (e.g. suitable areas within the study region) are at the bottom (Höfer et al., 2016; Stefanakou et al., 2019). At its simplest, this hierarchy, and thus tidal energy site suitability, is based on four criteria: resource feasibility/potential (red), environmental protection (green), economic uses (blue), and social uses (black). Several constraints on tidal energy feasibility were also identified, those being: peak spring tidal current speeds less than 1.5 m/s, depths less than 5

m and greater than 100 m, areas with existing tenures, and protected parks/areas (Cradden et al.,

2016; O'Rourke et al., 2010b; Segura et al., 2017b; Vazquez & Iglesias, 2016a). At this

screening stage 1.5 m/s was taken as a cautious and conservatively low threshold as areas should

not be screened out unless it is absolutely apparent that they would not be viable. In this sense,

those below 1.5 m/s are technically infeasible, while those areas not screened out do not

guarantee economic viability. Appendix B provides criteria and constraint rationale and

additional details.

The second step of the AHP is the development of a pair-wise comparison matrix of

evaluation criteria according to Saaty's nine-point scale (Table 2 below for Saaty's scale)

(Giamalaki & Tsoutsos, 2019; Höfer et al., 2016; Stefanakou et al., 2019). This is where

interviewees contributed to decision making by favouring the importance of each criteria relative

to one another within the matrix (Höfer et al., 2016; Saaty, 1980; Stefanakou et al., 2019).

Table 2: AHP pair-wise comparison scale (Saaty, 1980) taken from Wang et al., 2009.

Intensity of weight	Definition	Explanation
1	Equal importance	Two criteria contribute equally to objectives
3	Weak/moderate importance of one over another	Experience and judgement slightly favour one criterion over another
5	Essential or strong importance	Experience and judgement strongly favour one criterion over another
7	Very strong or demonstrated importance	A criterion is favoured very strongly over another; its dominance is demonstrated in practice
9	Absolute importance	The evidence favouring one criterion over the other is of the highest possible order of affirmation
2, 4, 6, 8	Intermediated values between the two adjacent scale values	Used to represent compromise between the priorities listed above
Reciprocals of above non-zero number		If criteria i has one of the above non-zero numbers assigned to it when compared to criteria j, then j has the reciprocal value when compared with criteria i

The third stage of the process focused on the elucidation of weights for each of the criteria (Stefanakou et al., 2019). This was achieved by: a) summing the values of each column within the pair-wise comparison matrix, b) normalizing the matrix by dividing each element of the table by the total sum of the column in which it belongs, and c) calculating the average of the elements in each row to determine each criteria's weighting (Mahdy & Bahaj, 2018; Stefanakou et al., 2019). The sum of all the criteria weights must be equal to 1.

The last stage of the AHP entailed a sensitivity analysis, achieved by calculating the Consistency Ratio (CR) (see Appendix B for methods) (Saaty, 1980; Stefanakou et al., 2019; Wang et al., 2009). The CR is used to validate the pairwise comparison assumptions and is essentially the ratio between the consistency of a given evaluation matrix and the consistency of a random matrix. It is an important step as it ensures that the judgements considered were consistent (Stefanakou et al., 2019; Wang et al., 2009).

At this point the interviews and the AHP had provided the required weightings for each criterion to create the Social and Economic uses sub-models as well as informing the Tidal Development Perspective weights.

4.2   Data overview

Data was taken from a variety sources including: the BC Marine Conservation Analysis (BCMCA) for economic, social, and environmental marine uses; Geo BC for base maps and provincial tenures; the Department of Fisheries and Oceans Canada (DFO) for catch values and volumes; the MaPP for regional and sub-regional MSP; and Natural Resources Canada (NRCan) for the peak spring current speed and bathymetry data. Appendix C details each dataset along with providing further background on sources.

4.2.1 Project set up

Each data layer used in this study was added to ArcMap 10.7.1 and pre-processed for use based on the general parameters described below, layer specific setup described in sections 4.3.1 - 4.3.4.

The *North American Datum (NAD) 1983 BC Environment Albers* was used as the Projected Coordinate System for this study, as it holds area constant and covers the entire study area. Furthermore, it is favored in mapping exercises by BC government ministries, while also being one of the more frequently used projections for the datasets used in this study (Ban & Alder, 2007).

For all raster layer creation, a cell size of 67 m x 67 m was chosen as it was representative of the smallest grid resolution of the tidal and bathymetric datasets provided by NRCan (Canadian West Coast Tidal Resource Assessment model). While the large range in the spatial resolution of the datasets would in turn yield the frequently encountered modifiable areal unit problem (MAUP), a focus on the most crucial dataset (e.g. resource feasibility) along with a cell size that would, to the greatest extent possible, preserve this dataset was chosen. Although tailoring the project resolution to the primary dataset does not 'solve' the MAUP, it does mitigate some of its effects (Dark & Bram, 2007).

4.3 Resource and marine use layers creation and normalization

Following pre-processing and layer creation, normalization of criteria to a common scale was required to enable meaningful comparisons and evaluation of trade-offs between uses represented by disparate data types. While the tidal current feasibility layer required the creation of its own distribution of suitability values (see 4.3.1), the remaining layers described in sections 4.3.2 - 4.3.4 were normalized using the formula:

$$Normalized\ layer = \frac{(X - X_{min})}{(X_{max} - X_{min})} \times 10 \qquad 4.1$$

Normalization ensured each criterion was comparable on a scale of one to ten and allowed for each to then be reclassified based on perceived compatibility with tidal energy sites. For instance, economic uses with a high normalized value (e.g. 10, representing significant use in each area) were reclassified as a 1 in terms of their contribution to site suitability, as per their optimization outlined in Appendix B. Figures for each layer described in sections 4.3.2 - 4.3.5 can be found in Appendix F.

4.3.1   Tidal stream turbine technical feasibility and constraint layers

To create the constraint layer, the tidal and bathymetric datasets had to be filtered to remove unfeasible locations, those being areas with peak spring tidal current speed less than 1.5 m/s and depth less than 5 m or greater than 100 m and clipped to the study region (Figure 7a below).

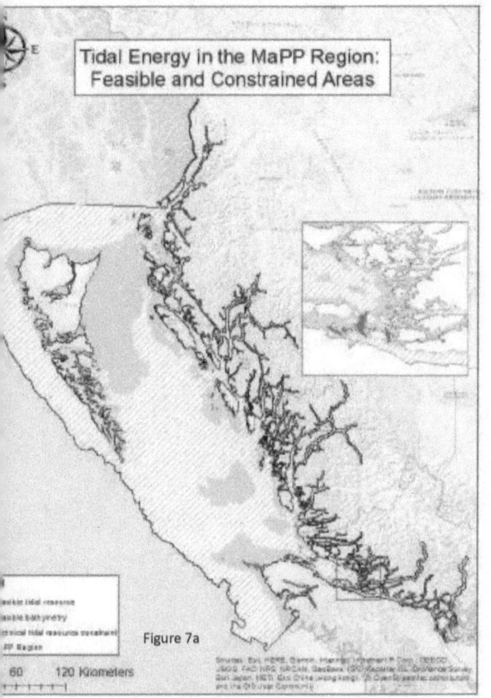

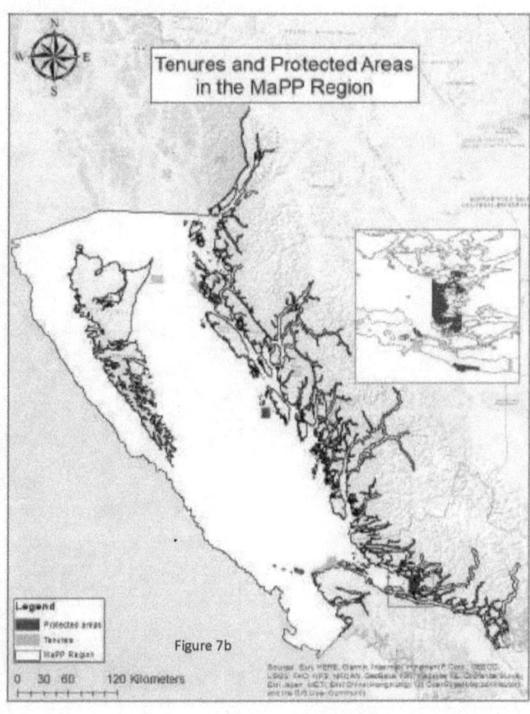

Figure 7: Technical feasible and constrained areas for tidal energy (a) and tenures and protected areas in the MaPP region (b).

This yielded a constraint layer in which tidal energy is deemed unfeasible from a resource availability and technical perspective, along with a feasibility layer representing sites that may be considered for tidal development. Next, the tenures dataset was filtered to remove reserve tenures which are identified and set aside by FLNRO for specific exclusive uses in the future and thus are not necessarily indicative of a use currently being there. The remaining tenures were combined with protected areas (ecological reserves, protected areas, and provincial parks) to create a regulatory constraint layer (Figure 7b) which was then removed from the feasible tidal energy area to create a layer of feasible depth and peak current constrained by existing tenures (Figure 8 below).

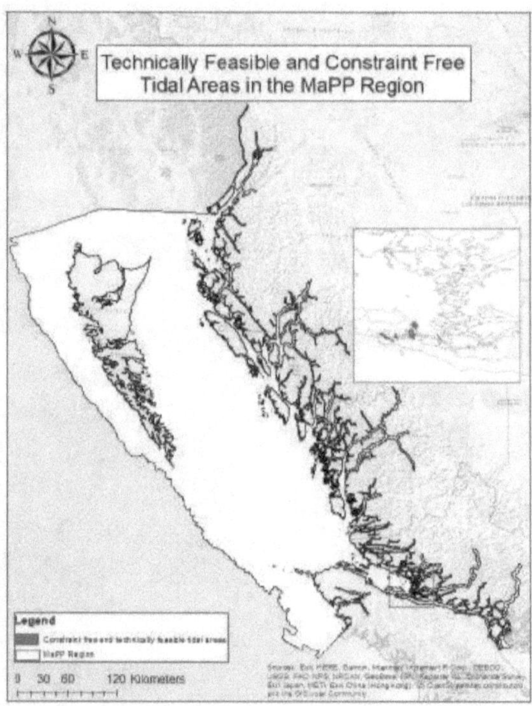

Figure 8: Technically feasible and constraint free areas for tidal energy development in the MaPP region.

4.3.2   Tidal Resource sub-model

While the literature often cites threshold values for suitable tidal currents, either based on peak or mean peak tidal current speed, there is no standard distribution of suitability for either metric. It was found that normalizing the data would skew the results as there were large differences in the maximum values for peak spring current speed across the study region, along with outliers being far removed from the mean, which would make tidal energy appear less suitable in the study area. Thus, a measure for converting peak spring current speed to the same suitability classification scale as the other criteria was needed.

Truelove (2020) developed dimensionless scaling models for wind, tidal stream, and wave energy based on performance data for actual technologies, which could be utilized to derive a suitability distribution. Truelove found that a site would be unconditionally non-viable, and thus representative of a lower bound on peak current speed suitability, if current speeds were less than the threshold determined by a fuel price input in Equation 4.2 below:

$$S < -0.0974F^2 - 0.8328F + 2.6847 \qquad 4.2$$

Where S is the annual peak tidal speed (m/s) and F is the diesel fuel price (CAD/L). To utilize Equation 4.2 an average fuel price for diesel in remote communities had to be determined (outlined in Appendix D). A fuel input of \$1.05/L was used in Equation 4.2 yielding a lower bound on peak current feasibility of $\approx$1.702 m/s.

An upper bound threshold for peak current speed could also be determined, assuming an installed capacity to annual peak electrical load capacity ratio of 1, by utilizing Truelove's unconditional viability formula:

$$\frac{P_r}{L} < 1.5 - 34.3193e^{-1.4063S} \qquad 4.3$$

Where P_r is installed turbine capacity (kW) and L is the annual peak load (kW). Assuming $P_r/L = 1$ yields an unconditionally viable annual peak spring current speed of $\approx$3.00 m/s. Treating these values as a lower threshold (i.e. suitability value of 1) and upper threshold (i.e. suitability value of 10) for peak current speed suitability, a trendline could then be applied to determine threshold current speeds for suitability values 2-9 (Figure 9 below).

As a TSTs performance is analogous to that of a wind turbine, it follows that a turbine reaches its rated capacity at the upper threshold for viability and would continue at this level of viability until its cut out current speed is reached. Cut out speeds for a range of turbines are summarized in Appendix D, either found through available data online or by rearranging and solving for current speed using Truelove's calculated mean dimensionless cut out speed of 0.65 and the equation:

$$|\hat{v}| = \frac{|v|-|v|^*}{|v|^*} \qquad 4.4$$

Where $|\hat{v}|$ is the mean dimensionless cut out speed, $|v|$ is the hub current velocity (taken as the cut out speed), and $|v|^*$ is the current speed at which rated performance is first achieved. This produced a mean cut out current speed of ≈5.26 m/s and a median cut out current speed of ≈5.03 m/s. Of the eight turbines, five had cut out speeds below the mean and thus the median value was used. From there, a linear trendline was drawn to the maximum cut out (6.5 m/s).

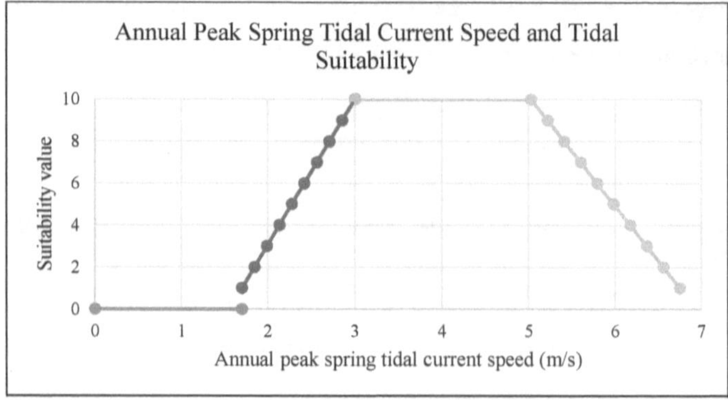

Figure 9: Piecewise function of annual peak spring tidal current speed and corresponding suitability value.

4.3.3 Economic Uses sub-model input layers

The Economic Uses sub-model consisted of six layers: commercial fisheries catch value, government/research vessel traffic, merchant vessel traffic, passenger and cruise vessel traffic, tug and service vessel traffic, and tanker traffic. See Appendix B for inclusion rationale. Aside from the government/research vessel traffic, which were combined as there is often overlap in researchers utilizing government vessels along with both often functioning in similar ways (e.g. environmental monitoring, sampling, and other data collection) each of the vessel traffic layers were indicative of a single sector.

Commercial fisheries catch value

The commercial fisheries catch value was calculated using methods derived from Xu (2018). This involved two data sources, BCMCA datasets providing spatial information on where historical catches have occurred and the annual average catch weight for species in those areas, while data sets from DFO provides both annual weight of commercial landings and value of commercial landings for species (BCMCA, 2011; DFO 2018). See Appendix C for a table of species.

For each fishing zone the total catch weight (w_c) was first divided by the time span (t) to calculate the annual landing (w_a) through equation 4.5. The annual landing was then divided by the area (A) of each fishing zone in order to estimate the annual landing density (D_l) in kg/km^2 through equation 4.6. Finally, the calculated annual landing density of each species was multiplied by its price (p) to calculate the annual landing value (V_l) using equation 4.7.

$$w_a = \frac{w_c}{t} \qquad 4.5$$

$$D_l = \frac{w_a}{A} \qquad 4.6$$

$$V_l = D_l \times p \qquad 4.7$$

The landing values for each species subset were combined using the *Field Calculator* and *Cell Statistics* tools in ArcMap. To reduce the effect of large and small outliers in the data, the 97th and 3rd percentiles were calculated and removed (i.e. values above and below the 97th and 3rd percentile were reclassified as the 97th and 3rd percentile values, respectively). While this did create a two 'walls' of the same values (high and low), it succeeded in reducing the effect of outliers. See Appendix E for the histogram distributions of each layer. The *Polygon to raster* tool was then used with a cell size of 67 m x 67 m to match the tidal resource feasibility layer.

Marine vessel traffic

The Marine Vessel Traffic Data taken from the BCMCA illustrates the density of marine vessel traffic in hours transiting within 5 km x 5 km grid cells in Canadian Pacific waters for 2010. Traffic is categorized based on vessel type: fishing, government, merchant, passenger & cruise, pleasure and yacht, research, tanker, and tug & service vessels being available subsets. The Canadian Coast Guard-Marine Communications and Traffic Services vessel tracking database provided the datasets, which were processed by the Dalhousie University Maritime Activity and Risk Investigation Network.

For this analysis, the fishing vessel traffic dataset was removed, as commercial fishing catch value was already representing this use and thus would go against the criteria consistency principle outline in section 4.1.1. The importance of each vessel type used and rationale for inclusion can be found in Appendix B. The vessel subsets were filtered to reclassify the 95th and 5th percentile as threshold values to reduce the impact of outliers on the data.

4.3.4   Social Uses sub-model input layers

The Social Uses sub-model consisted of three datasets: pleasure and yacht vessel traffic density, sport fishing feature count, and recreation and tourism feature count.

The pleasure and yacht traffic layer represent recreational boating in the study region for vessels greater than 20 m in length. While this is potentially poor representative of overall recreational boating traffic as it lacks information regarding the spatial distribution and density of smaller vessels, it represents the most tangible data set spanning the study region. This dataset was also filtered to reclassify the 95th and 5th percentile values.

Sport or recreational fishing is an important social and tourism activity. The sport fishing feature count illustrates areas in which effort is expended in each 2 km x 2 km grid, not necessarily where fish are caught, for four fisheries: anadromous fish, crab, groundfish, and prawn and shrimp. The data reflects both independent and guided participation, sourced from the DFO Crab Fishery distribution on the North Coast 2007 (updated by the BCMCA with local knowledge), Parks Canada, and the BC Coastal Resource Information Management System (updated by the BCMCA with local knowledge).

Tourism and recreation features provide an indication of social uses within the study area, while not a perfect representation, it does provide a proxy of social uses in the study region. The BC tenures dataset was used to create the tourism and recreation feature layer based on the BCMCA Tourism and Recreation Feature Count dataset. Tenures were filtered to remove reserve tenures and selected based on tenure purpose being one of: residential, community, commercial recreation, or First Nations; and/or tenure sub-purpose being one of: local/regional park, school, marina, private yacht club, or public wharf. A 2 km x 2 km grid (matching the resolution of the BCMCA dataset) was created using the *Create Fishnet* tool which was then used to determine how many recreation and tourism features intersected each grid cell.

4.3.5 Marine Conservation Value sub-model

Ecological services and ecosystem integrity are crucial not only for the species that inhabit the region, but also for economic activities, recreation, and socio-cultural values imbued in nature. The BCMCA Ecological Marxan results Expert High (no clumps) dataset was used as a proxy representing ecological value within the study region. The Marxan spatial analysis was undertaken to identify areas of high conservation value using available ecological data, conservation targets, and recommendations gathered from 'expert' workshops. The Marxan scenario provides normalized data on a 0-100 scale, 0 being areas of low conservation value and 100 being areas of high conservation value which were reclassified to a 1-10 scale.

4.4   Phase I Weighted Overlay sub-model and model creation

The Economic and Social Uses sub-models were created for each MaPP sub-region by combining the input layers based on their AHP derived weightings using the *Weighted Overlay* tool. The *Weighted Overlay* tool applies the following formula to the raster cells of each input layer:

$$Suitability = \sum_{i=1}^{n} W_i X_i \qquad 4.8$$

Where W_i is the weighting for the ith criteria and X_i is the value of that criteria (Mahdy & Bahaj, 2018).

Within the Social Uses sub-model, the tourism & recreation and sportfishing features layers were given no data values of 8 to ensure spatial coverage of each layer across sub-regions, and thus a contribution to site suitability. This was chosen based on the assumption that the absence of data is not indicative of there being no use, nor is indicative that the area may be highly constrained.

The Economic and Social Uses sub-model weights were taken as the average weights for each criterion from the co-lead AHP interview responses for each sub-region. As co-leads for the North Coast were unable to participate, weights were taken as the combined averages of the FLNRO and FN co-lead responses respectively.

For each sub region, the sub-models (Tidal Resource, Social Uses, Economic Uses, and Marine Conservation Value) were combined using the co-leads average weights derived from the Tidal Development Perspective model (Table 7). An Equal Weights model was also created to examine the sensitivity of outputs to input alterations. The processing extent was set to the technically feasible areas (refer to Figure 8) to avoid both regulatory and technical constraints.

The Tidal Development Perspective model weights were based on the rank order formula:

$$Weighted\ sum = \frac{2(n + 1 - r)}{n(n + 1)} \qquad 4.9$$

Where n is the number of factors and r is the ranking of each factor (Stillwell et al., 1981). The derived weighting for each ranking can be found in Table 3 below.

Table 3: Sub-model ranks and equivalent weighting.

Rank	Weighting
1	40%
2	30%
3	20%
4	10%

Chapter V: Phase I results

5.1   Phase I interviews

All interviews were conducted over the phone between January-April 2020. Each participant was asked the questions depicted in Figure 10 below.

1. Rank the following on a Likert scale of 1-5 (1 being unimportant, 2 being somewhat important, 3 being important, 4 being very important, and 5 being extremely important) the importance of each of the following drivers for the electrification of remote diesel reliant communities:
   - A)   Energy self sufficiency
   - B)   Climate change mitigation
   - C)   Providing opportunities for economic development
   - D)   Providing opportunities for community infrastructure development/upgrades
   - E)   Reducing electricity costs
   - F)   Avoiding health impacts associated with diesel generation
   - G)   Avoiding environmental impacts associated with diesel (risk of spill, ground contamination, etc.)
   - H)   Other(s), please specify

2. Rank the following on the same Likert scale of 1-5 as barriers to the electrification of remote diesel reliant coastal communities:
   - A)   Funding
   - B)   Federal policies
   - C)   Provincial policies
   - D)   Information to assess and determine renewable options
   - E)   Other(s), please specify

3. When considering the addition of a new use of the marine space, such as tidal energy, what characteristics (i.e. spatial footprint, compatibility with other uses, environmental impact etc.) of the use are paramount from the perspective of the (*insert*) Nation to consider?

4. Are you familiar with or do you have knowledge of the following characteristics of tidal energy? Please answer on a Likert scale of 1-5.
   - A)   How the technology works
   - B)   Spatial requirements
   - C)   Environmental impacts
   - D)   Regulatory requirements
   - E)   Cost of electricity

5. Fill out the Analytic Hierarchy Process matrix below based on the relative importance of each economic use pairwise comparison when considering the suitability of tidal energy.

6. Fill out the Analytic Hierarchy Process matrix below based on the relative importance of each social use pairwise comparison when considering the suitability of tidal energy.

7. Rank each of the following uses for each perspective. Note that the theme of each perspective is ranked 1 in each case, and tidal resource is ranked at least second as this exercise is fundamentally based on identifying suitable locations for tidal energy development in proximity to remote coastal diesel reliant communities.

8. What concerns, if any, do you have with the potential development of tidal energy in the MaPP region to replace diesel generation in remote coastal communities from a Marine Spatial Planning perspective?

Figure 10: Phase I interview questions. See Appendix G for tables related to questions 5-7.

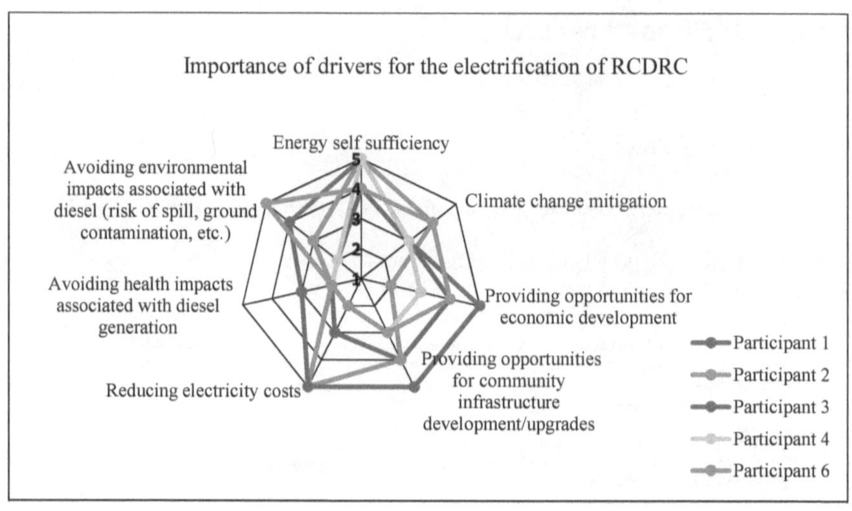

Figure 11: Comparison of responses to the importance of drivers for the electrification of First Nations RCDRC.

Figure 11 shows participant responses to question 1, with energy self sufficiency (average 4.6), providing opportunities for community development (3.8), and economic opportunities (3.6) having the highest average rankings. When asked if there was anything else participants would like to include, several additions were made. Participant 2 added to energy self sufficiency (A) the need for reliable backup power supply in case of power outages, while participant 5 noted the difficulties of navigating BC Hydro's power monopoly, stemming not only from the crown corporations overarching control, but also in the ways power is produced. Furthermore, they described past and current issues surrounding respect for and recognition of Indigenous rights, especially in instances of treaty rights being violated, citing Site C as an example. Participant 3 added that within the Central Coast sub-region there are areas that had been considered for tidal energy (with one SMZ created specifically for tidal) but that most of the interest was not in the electrification of remote communities, but in economic development opportunities. Going on to describe interest in the role of tidal energy in powering a remote operation, such as shellfish

aquaculture, as a major barrier to economic development on the coast is the availability of power.

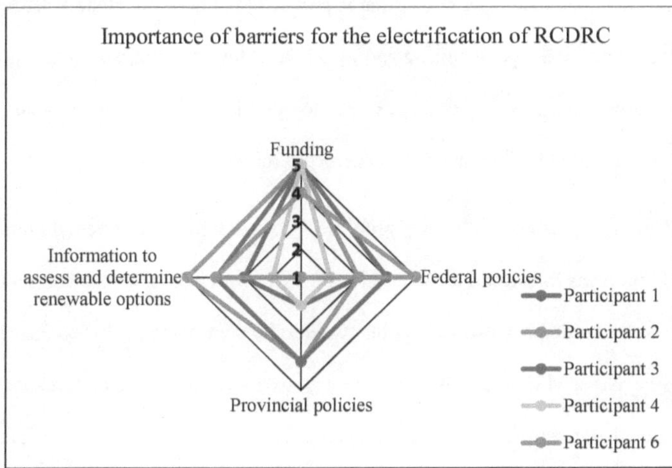

Figure 12: Comparison of responses to the importance of barriers to the electrification of remote diesel reliant coastal communities.

In terms of barriers (Figure 12), funding was by far identified as the most important with

an average score of 4.8. The importance of information to assess renewable options in RCDRC

was the second greatest perceived barrier with an average score of 3.6 while also having the

greatest range in responses (1-5). Participant 6 described the importance of obtaining information

on energy options, along with illustrating that information is needed not only on the engineering

and technical feasibility side, but social and cultural information to inform energy decision

making. Government policies at both the federal and provincial levels were viewed as important

and somewhat important, respectively. Participant 6 noted the importance of federal policies

while also describing the challenges of trying to work across multiple government agencies and

subsequent funding barriers. When asked if they had anything else, Participant 2 added "The capacity to pursue funding, policy opportunities, do assessments, and understand how assessments apply to a particular community" and viewed it as being very important. Participant 3 added the challenge of storing power that is produced but not immediately used. Participant 6 also described the challenge of storing energy while adding: "I think the status quo is also a barrier. What we have currently in place even though it is diesel and can be quite challenging, but it is a cheap form of electricity and it has been very reliable".

Regarding question 3, the most prominent answer was environmental impact closely followed by impacts to cultural uses and historic sites, with respondents either mentioning one of the two or both. Participant 1 described that when tidal areas were identified based on resource presence in existing MSP, the primary concerns raised were related to cultural uses which had not been included in what had been a spatial analysis focused on technical feasibility. Other questions arose the required resolving such as information related to device characteristics (such as durability and longevity), which company to choose, required infrastructure, benefits to their community and others from tidal, and how a device would feed into the grid. For participant 6 the most important thing to consider is the acceptable type of electricity source on Haida Gwaii. Furthermore, they remarked that encompassing spatial planning exercises to identify energy technology suitability and compatibility with other uses and values has not been done on Haida Gwaii as of yet, while stressing the importance and utility of undertaking such endeavours.

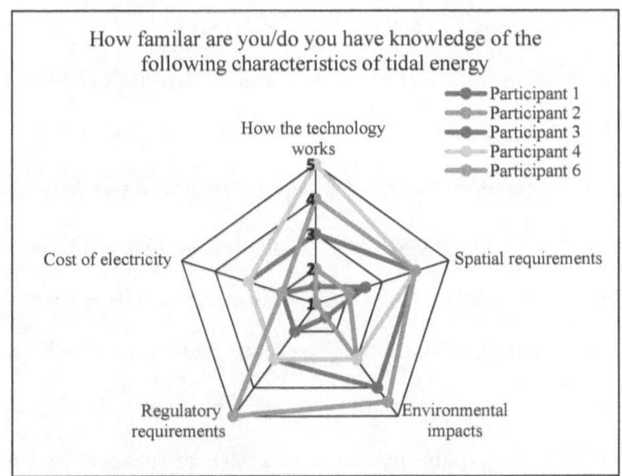

Figure 13: Assessing participant familiarity with tidal energy characteristics.

Figure 13 above gauged participants knowledge/understanding of tidal energy

characteristics and illustrates that there is still much uncertainty surrounding the technology.

While the spatial requirements of devices received the highest average score (3.3), this is still an

average response of "unsure". Participant 1 noted that although they had identified areas of tidal

energy potential, they were still unsure of what an actual device would look like in the water and

how it would produce electricity, citing the need for more research and provision of information

to communities. Participants were also unsure of the environmental impacts from devices.

Participant 1 elaborated on his response (strongly disagree/disagree), highlighting

misconceptions around the environmental impact of devices and how, at this point, an

understanding of environmental impacts just is not there as they haven't been privy to any new

research. On average participants were also unsure of how the technology works (3.1) and the

regulatory requirements associated with tidal energy (2.8). The least understood characteristic of

tidal energy was the cost of electricity with an average response of 2.2 (respondents disagree that they are familiar) with average responses of 1.66 and 3 for FN and FLNRO respectively. For each question, FLNRO participants had a greater average perceived understanding (even if they were unsure) of tidal energy than FN participants, pointing to potential information/data accessibility disparities between the two. Participant 1 described that there was input from the tidal sector earlier in the MaPP process and even discussions of establishing a MRE research centre near Campbell River similar to FORCE in Nova Scotia, however nothing really ever became apparent. As a result of this from the perspective of the member FN communities his organization represents, there is not a solid understanding of the technology as there has been few examples of technology testing or tried and true technologies deployed in the region.

Government and research vessel traffic had the second highest weighting and also the largest range in weightings (8-36, see Table 4 below) within the Economic Uses sub-model, potentially indicating some uncertainty regarding the importance of the layer. Tanker vessel traffic was perceived as having the lowest potential conflict with tidal energy development, though this may have also been a function of tanker traffics importance within the MSP for each group as there are several First Nations that oppose tanker traffic in their traditional territories. The consistency ratio values were quite high, with participants 3 and 4 being the only two that did not exceed the suggested threshold for consistency ratios. Participant 4 observed the inherent difficulties of providing weightings for entire MaPP sub-regions, in that the spatial distribution of activities is not fixed and thus different weightings are applicable in different areas. See Appendix G for AHP matrices and section 4.1.1 for AHP methods.

Table 4: Weightings for each economic use derived from the AHP and consistency ratio. The sum of each participants weightings is equal to 100.

	Participant 1	Participant 2	Participant 3	Participant 4	Participant 6	Average weight
Commercial fisheries catch value	41	44	30	49	19	36.6
Government / research vessel traffic	25	10	20	8	36	19.8
Merchant vessel traffic	13	6	16	15	2	10.4
Passenger and cruise vessel traffic	6	24	12	11	18	14.2
Tanker vessel traffic	2	2	11	2	3	4
Tug and service vessel traffic	13	14	11	15	22	15
Consistency ratio	0.175	0.175	0.126	0.137	0.233	N/A

The tourism and recreation feature count layer received by far the most weighting within the Social Uses sub-model, doubling the second highest (pleasure and yacht traffic, see Table 5 below). Sport fishing and pleasure and yacht traffic had similar weightings, but it is clear that the broader and more encompassing measure of social activities is perceived to have a greater spatial overlap with tidal energy than the more recreational activities examined individually.

Table 5: Weightings for each social use derived from the AHP and consistency ratio.

	Participant 1	Participant 2	Participant 3	Participant 4	Participant 6	Average weight
Sport fishing feature count	9	26	20	33	6	18.8
Pleasure craft and yacht traffic	21	11	20	33	30	23
Tourism and recreation feature count	70	63	60	33	64	58
Consistency ratio	0.028	0.033	0	0	0.091	

Question 7 was intended to derive weightings for each of the sub-model inputs within the Tidal Development Perspective model; along with economic, marine conservation, and social perspectives. However, a discussion with participant 2 highlighted that the formatting of the question limited and constrained the ability of respondents to enact their organizational MSP. Thus, in keeping with this studies goal of being holistic and receptive to stakeholder feedback, it was decided to only focus rank weights for the Tidal Development perspective (e.g. the model for determining suitable locations in Phase II). See Table 6 below for ranks.

Table 6: Tidal Development Perspective sub-model input ranks.

Sub-model	Participant 1	Participant 2	Participant 3	Participant 4	Participant 6
Tidal resource suitability	1	1	1	1	1
Economic Uses suitability	3	4	2	3	4
Social Uses suitability	4	3	4	2	3
Marine Conservation Value suitability	2	2	3	4	2

The final interview question was intended to assess any concerns participants might have with the potential for tidal development in the MaPP region from an MSP perspective. Participant 1 had concerns with the lack of progress in terms of tidal development within the NVI region, stating that it may be due to a lack of funding and political interest but also that there hasn't been anyone who has really "championed" it and driven development.

Participant 2 was concerned with ecological conservation and the preservation of cultural values, along with questions regarding the capacity to implement tidal projects and the longevity of such devices, stating: "The actual feasibility of running a complex technology in a remote community is definitely a lens that we would look through at something like this". Another concern they had was community familiarity with diesel raising a suite of issues in terms of replacing infrastructure with RETs, such as whether a community has the human resources (HR) capacity to do so, along with the need to assess/compare a range of renewable options (beyond tidal).

First and foremost, participant 3 had concerns with the potential environmental impacts from tidal devices. A process which they described would first require the identification of potential sites and then the evaluation of environmental values at that location. Secondly, ensuring potential sites do not infringe upon traditional First Nation's harvests or cultural values along with having the support of FN. They also described the importance of examining the proximity of sites to protected areas, potential impacts to recreation activities, and noted the potential for tidal energy development to support economic development.

Participant 4 noted that Haida Gwaii and FLNRO welcome the opportunities that tidal energy could bring, along with the reinforcing the need for identifying feasible locations and the need for cultural and ecological impacts to be closely examined and mitigated.

Participant 5's main concern entailed the potential for tidal energy to be developed without the full participation/consent from First Nations communities and without benefit to those communities. Furthermore, the potential for companies to push products on communities who have limited capacity or might be taken advantage of to test their technologies. They also described the need to build community capacity from the get-go through the development of training programs to educate people beyond the basics of tidal energy. Finally, they stated the need for support from both the federal and provincial government to help build this capacity.

Participant 6's primary concern is that the exercises to identify suitable sites have not been undertaken yet while also stating the importance of doing so, and thus other RETs (such as solar) are being pursued.

5.2 Phase I GIS-MCDA sensitivity analyses

5.2.1 Sub-model sensitivity analysis

Sensitivity analyses are a critical component of any modelling study. Sensitivity analysis allows researchers to analyze the contribution of different inputs (e.g. criteria) to model outputs and identify sources of variation to model results (Kocabas & Dragicevic, 2006). Overall, the Economic Uses sub-model inputs appear to have minor impacts on the percent areal coverage of different suitability levels, with the merchant vessel traffic layer yielding the greatest change. The alterations to the Social Uses sub-model inputs had a much greater impact on percent areal coverage results than the Economic Uses sub-model input layers. The sensitivity analyses for each sub-region can be found within Appendix G.2.

5.2.2 Model sensitivity analysis

A multivariate sensitivity analysis was used to examine the sensitivity of the Tidal Development Perspective suitability model to variations in sub-model weights. This was achieved by varying weights through an Equal Weights model and the Tidal Development perspective model informed by the rank weights derived from the Phase I AHP interviews. See Appendix H for model sensitivity analysis.

5.3   Phase I GIS-MCDA results

The small-scale maps in Figures 14-17 illustrate site suitability across each sub-region, while the large-scale maps in each represent suitability in a smaller area of the region. Refer to Appendix I for figures illustrating sub-regional sub models along with Equal Weights results.

5.3.1   North Vancouver Island sub-region

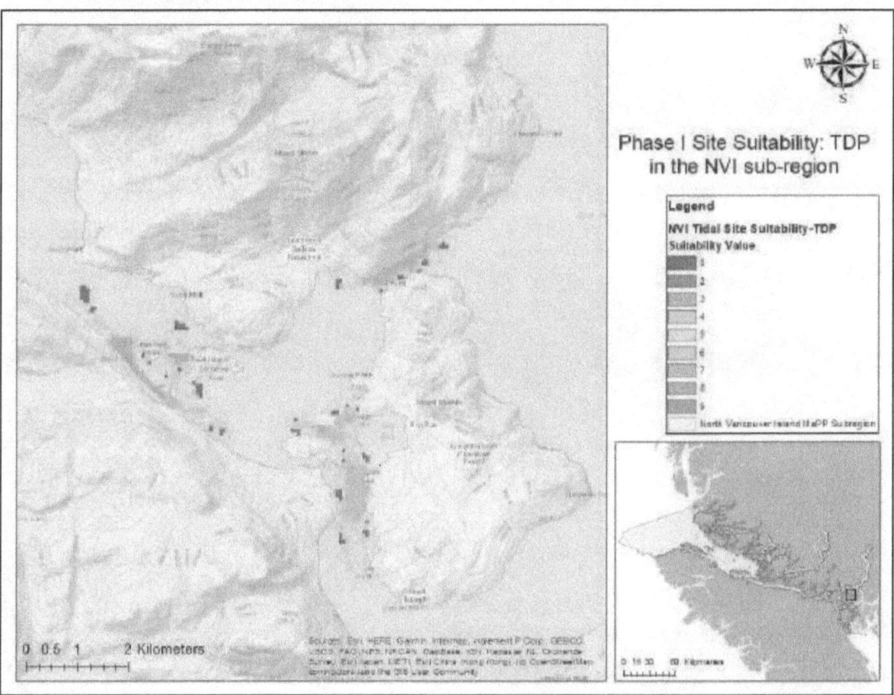

Figure 14: Tidal site suitability according to the Tidal Development Perspective scenario in the NVI sub-region. Large scale map illustrates suitability near Sonora and Stuart Islands (northeast of Campbell River).

The NVI region had by far the greatest area of potential tidal sites (suitability value > 0), spanning ≈61.51 km² (Figure 14 above). The Tidal Development Perspective model inputs were weighted as: Tidal resource suitability (TRS) 40%, Economic Uses suitability (EU) 25%, Social Uses suitability (SU) 10%, and Marine Conservation Value suitability (MCV) 25%. The suitable area is dispersed across many geographically distinct sites while also being the only sub-region containing a suitability value higher than 7.

5.3.2   Central Coast sub-region

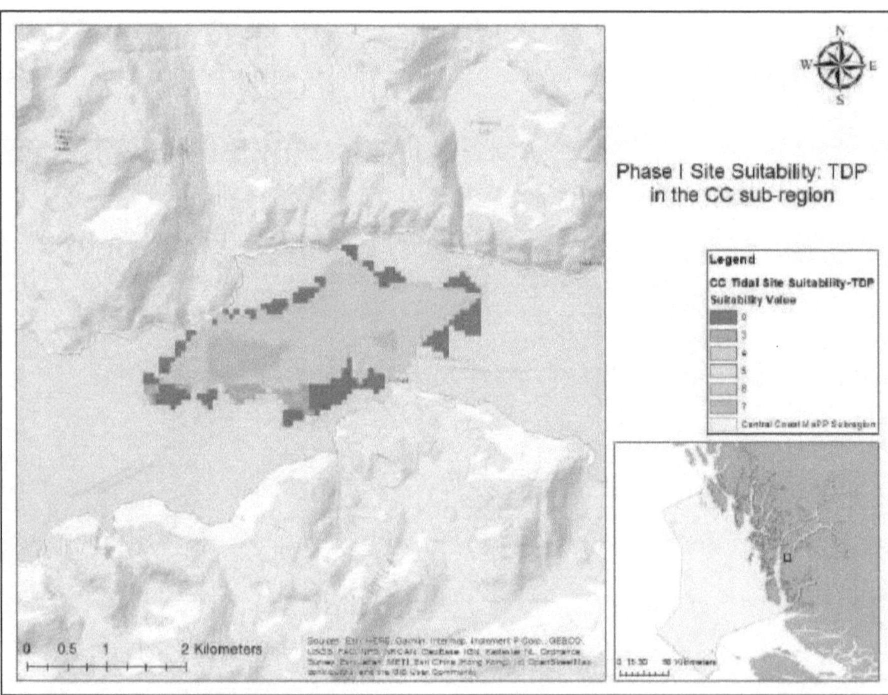

Figure 15: Tidal site suitability according to the Tidal Development Perspective scenario model in the Central Coast sub region. Large scale map shows site suitability in Burke Channel (southwest of Bella Coola).

The CC sub-region had the second smallest areal coverage by sub-region of potential tidal sites ($\approx$3.79 km^2), predominantly with a suitability value of 6 ($\approx$2.93 km^2) (Figure 15 above).

The Central Coast Tidal Development Perspective model inputs were weighted as: TRS 40%, EU 20%, SU 15%, and MCV suitability 25%. Potential area was primarily confined to one site in Burke Channel.

5.3.3   North Coast sub-region

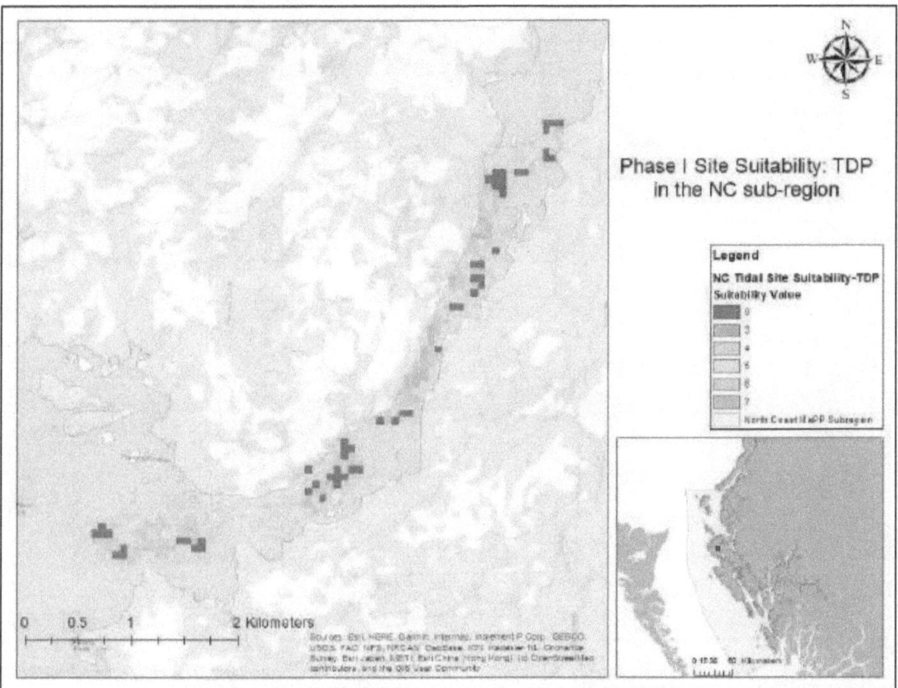

Figure 16: Tidal site suitability according to the Tidal Development Perspective scenario model in the North Coast sub-region. Large scale map shows site suitability in Porcher Inlet (south of Prince Rupert).

The NC sub-region had the smallest areal coverage (≈ 2.99 km^2) of potential tidal sites, being more dispersed than the clustered 'hotspot' observed in the CC sub-region (see Figure 16 above). The Tidal Development Perspective model inputs were weighted as: TRS 40%, EU 18%, SU 18%, and MCV 24%.

5.3.4   Haida Gwaii sub-region

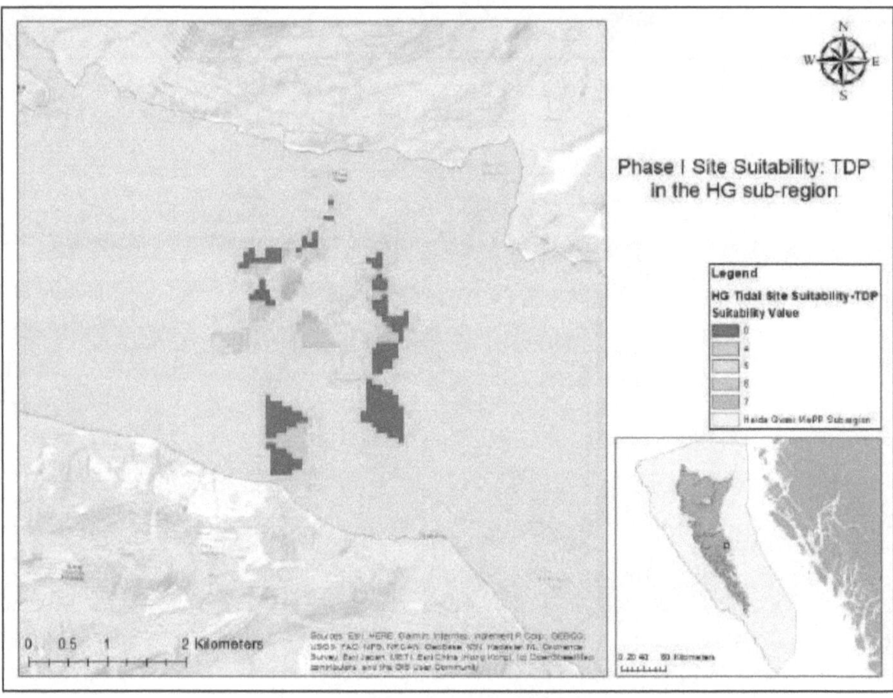

Figure 17: Tidal site suitability according to the Tidal Development Perspective scenario model in the Haida Gwaii sub-region. Large scale map shows site suitability in Cumshewa Inlet (south of Sandspit).

Finally, the HG sub-region boasted ≈ 20.02 km^2 of potentially suitable tidal sites. These sites were geographically dispersed across the north, east (e.g. Cumshewa Inlet, displayed in Figure 17 above), and south ends of Haida Gwaii. The Tidal Development Perspective model inputs were weighted as: TRS 40%, EU 15%, SU 25%, and MCV 20%.

Chapter VI: Phase II methods

6.1 Including economic site considerations and constraints for a candidate site

Building upon the Phase I GIS-MCDA, Phase II mapping was intended to identify a potential tidal site for a candidate community by incorporating supplementary feasibility considerations. These included economic factors; such as the distance from site to port (representing installation and maintenance travel costs), distance from site to communities (representing cabling connection costs), and depth (representing costs associated with the installation and maintenance of a TST device), along with the MaPP zoning for each potential site (recognising that a project is more likely to be considered or approved if it fits within existing MSP), and suitability values derived from Phase I MCDA-GIS (following the logic that the most suitable sites should be investigated first).

These additional metrics were included as the costs associated with developing a site are highly correlated with water depth, distance to the shoreline, suitability of the resource, and travel distance from installation and service ports (Dalton et al., 2015; Hemer et al., 2018; Kerr et al., 2015; Kilcher et al., 2016; Macdougall et al., 2013; O'Rourke, 2010a; Sangiuliano, 2017b; Segura et al., 2017; Van Cleve et al., 2013; Vazquez & Iglesias, 2016).

6.1.1 Site distance from ports

The Geo BC Port dataset was used to identify potential ports. This dataset was screened based on criteria from Nova Scotia's *Marine Renewable Energy Infrastructure Assessment* (see Appendix J). After applying this screening two major deepwater ports were identified: The Port of Vancouver and the Port of Prince Rupert.

The suitability of distances from deepwater ports (Table 7 below) was taken from the Pacific Northwest National Laboratories (PNNL) *Geospatial Analysis of Technical and Economic Suitability for Renewable Ocean Energy Development on Washington's Outer Coast* and converted from nautical miles to kilometers (Van Cleve et al., 2013). This was taken as the reference for suitability as it focuses on a similarly expansive study region, along with being at the same level of preliminary analysis.

Table 7: Site distance from a deepwater port and suitability value.

Suitability value	Distance from site to deepwater port (km)
10	<9.26
9	9.27-18.52
8	18.53-37.04
7	37.05-55.56
6	55.57-74.08
5	74.09-92.6
4	92.61-185.2
3	185.21-277.8
2	277.81-370.4
1	370.41 - 400

The *Euclidean Distance* tool was then used to create a mask illustrating distance derived suitability for all sites within 400 km of a major deep-water port (see Figure 18a). While this is not a perfect approximation of distance from port to site for a vessel, it does represent straight line distance "as the crow flies", while being a simpler more straightforward method than a cost-distance pathway analysis which was beyond the scope of this book.

6.1.2   Site distance from communities

Communities were pulled from the NRCan Remote Communities Energy Database (Appendix C). The dataset was screened to remove communities outside of the MaPP region and further screened to remove: No water access, grid connected, community record status = not

active, community type =commercial, and those with no population data. Further, those that were

not within 20 km of a technically feasible and constraint free tidal resource were also removed.

Distance from potential sites to the closest community water access was approximated as

distance from community to site. Distance suitability was derived from the PNNL report and the

distribution of values can be found below in Table 8.

Table 8: Site distance from community and suitability value.

Suitability value	Distance from site to shore/community (km)
10	< 1.85
9	1.86 - 3.7
8	3.71-5.55
7	5.56-7.4
6	7.41-9.25
5	9.26-11.11
4	11.12-12.96
3	12.97-14.81
2	14.82-16.65
1	> 16.66

The *Euclidean Distance* tool was again used to generate a distance suitability mask from

sites to communities (Figure 18b below). This did not consider several important factors

including route topography and whether the travel surface was terrestrial or marine.

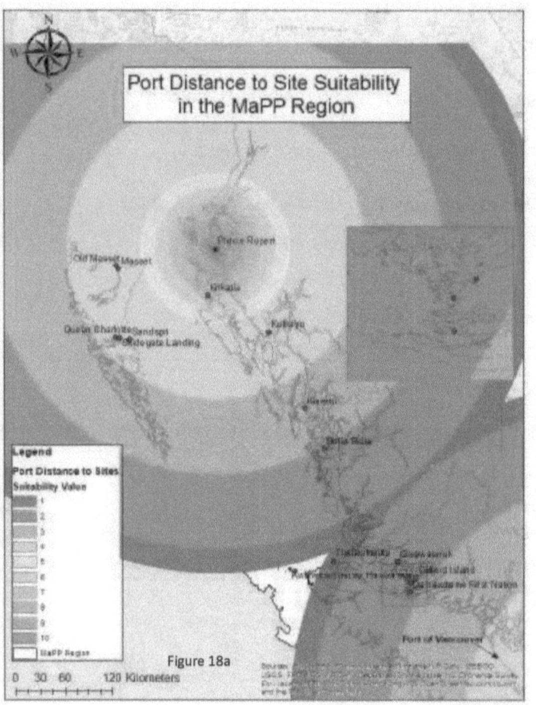

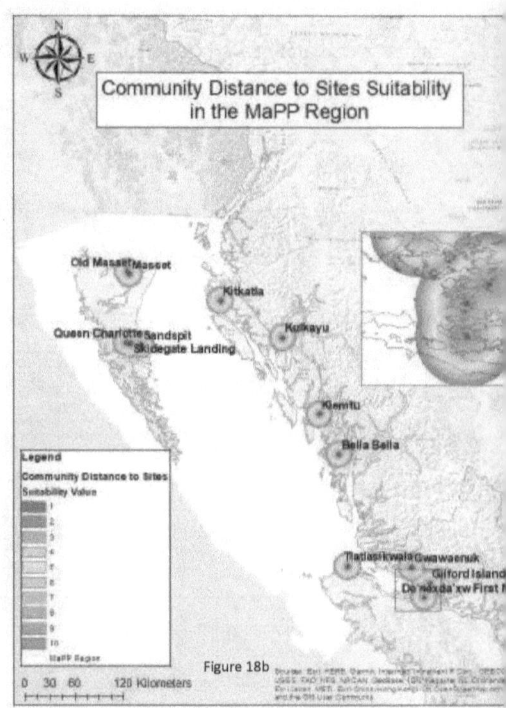

Figure 18: Site distance suitability in the MaPP region, from ports (a) and RCDRC (b).

6.1.3   Site depth

While the PNNL report did include suitability values across a range of depths, the report was focused on submerged devices and was thus altered to focus on the generic TST assessed in this study using a piecewise function modified from Kilcher et al. (2016). Whereas Kilcher et al. used a maximum depth of 150 m, this study used a maximum depth of 100 m, in keeping with the technical feasibility parameters outlined in section 4.3.1. The piecewise function is shown below in Figure 19 and defined as: sites shallower than 5 m are not technically feasible, sites between 20 and 60 m are ideal (suitability = 10), and sites deeper than 100 m are considered

unfeasible at this time. Suitability values change linearly between 5-20 m and 60-100 m.

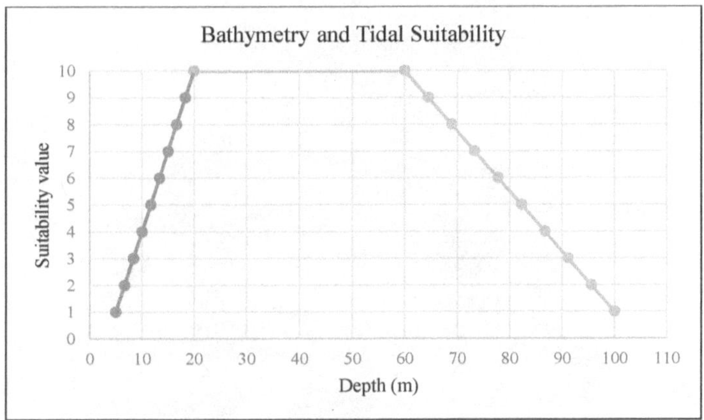

Figure 19: Piecewise function of bathymetry and corresponding suitability value.

The bathymetry suitability layer (Figure 7a) was converted to a rounded integer raster for ease of re-classification using the *Raster Calculator,* and then reclassified according to Table 9 below to create Figure 20.

Table 9: Depth and corresponding suitability value, derived from Figure 19.

Suitability value	Depth (m)
10	20-60
9	18-19, 61-64
8	17, 65-69
7	15-16, 70-73
6	13-14, 74-78
5	12, 79-82
4	10-11, 83-87
3	8-9, 88-91
2	7, 92-96
1	5-6, 97-100

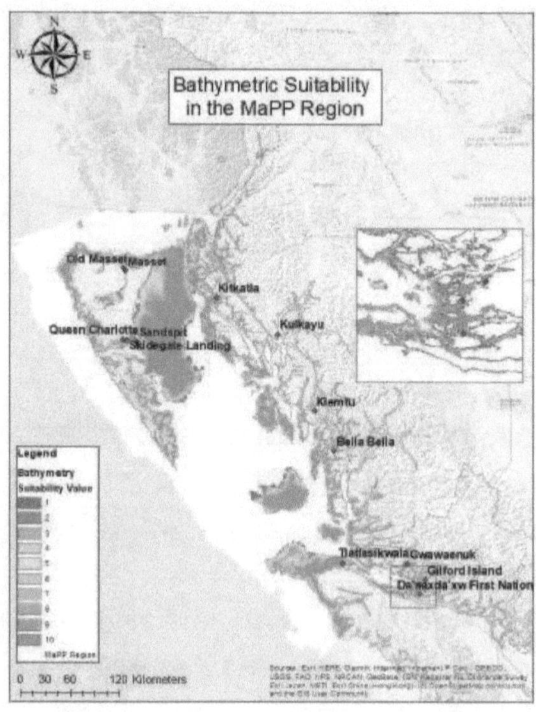

Figure 20: Bathymetric suitability for potential tidal sites in the MaPP region.

6.1.4 Site MSP zoning

GMZ and SMZs with a focus on alternative energy are the most suitable zoning schemes for tidal energy development. PMZs and SMZs in which alternative energy development is not allowed are the least suitable, while PMZs and SMZs in which alternative energy is conditionally allowed represent mid-level suitability. All being equal, sites with more favourable zoning for alternative energy development are favoured. See Figure 21 below for MaPP zoning.

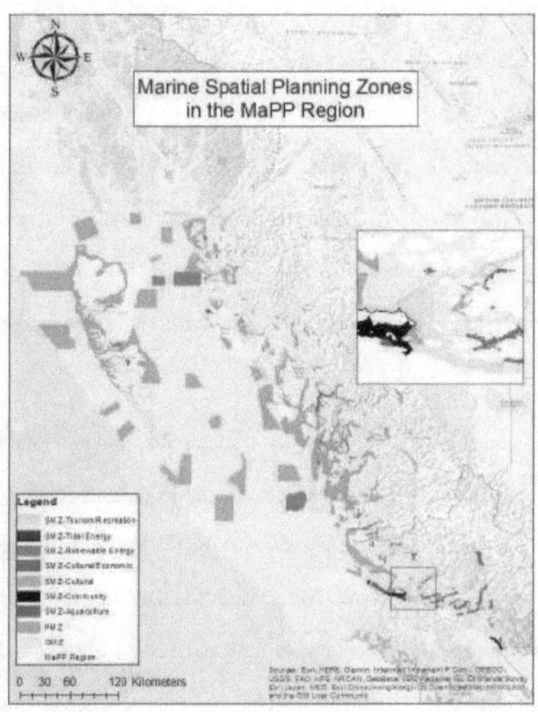

Figure 21: MaPP MSP zones. Data taken from http://mappocean.org/.

6.1.5   Phase I Tidal Development Perspective suitability model

Phase I site suitability values were also considered in the Phase II analysis based on the logic that sites with higher values are more suitable. The TDP suitability model for each MaPP sub-region was combined using the *Mosaic to New Raster* tool to allow for an encompassing assessment of all sub-region communities.

6.2   Phase II weighted overlay

The Phase II layers described previously were combined using the *Weighted Overlay* tool, with a similar weight distribution as the PNNL report: 45% TDP suitability model, 30% site distance to community suitability, 15% depth suitability, and 10% site distance to deepwater port suitability (Van Cleve et al., 2013). An equal weights scenario, along with scenarios in which depth, community distance, and port distance were each weighted as 40% (with the remaining criteria each given a weighting of 20%), examined the sensitivity of the model to the alteration of weights.

Sites were examined based on highest cell values. An approximation of the size requirements for a single device was derived from AECOM Canada's report for the Bay of Fundy which assumed an areal requirement of 500 m^2 for 1-3 MW of capacity, with an average generating capacity between 100-500 kW (AECOM, 2014). Dividing the maximum and minimum numbers of turbines for each bound (1 MW and 3 MW, 100 kW and 500 kW) yielded a range of single turbine areal requirements of $\approx$16.6 m^2-250 m^2, with a median value of 66.5 m^2. This value matches well with the cell size of this analysis and thus sites were investigated based on the highest cell value present where the number of high value cells is greater than two.

6.3   Candidate community interview

The Phase II candidate community interview had four primary goals: 1. Examine how existing diesel generators are viewed within the community 2. Determine the LCOE for the community (or obtain the necessary data to calculate an LCOE) 3. Assess the desired characteristics of future energy use community development 4. Gain feedback on the identified site and examine ways in which the mapping could be more representative.

Chapter VII: Phase II results

7.1   Phase II weighted overlay

Figure 22 illustrates site suitability near the community of Kitkatla along with total site suitability across the MaPP region, with a spatial coverage of 21.9 km^2 ($\approx$0.02% of the study area) and 15.30 km^2 being unsuitable (0). See Appendix K for the Equal Weights scenario results.

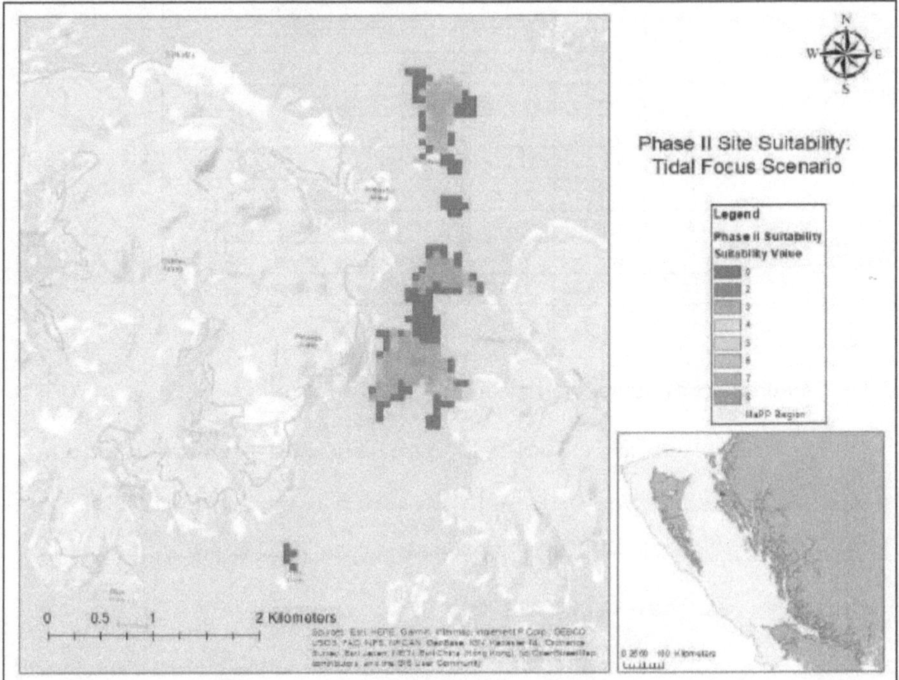

Figure 22: Phase II Tidal Focus Scenario. Small scale map illustrates Phase II suitability the community of Kitkatla (south of Prince Rupert) and the large scale displays Phase II site suitability across the MaPP region.

7.1.1 Phase II mapping sensitivity analysis

Overall, the Phase II multivariate sensitivity analysis (see Table 10 below) shows large variations in model output with alterations to input weights, especially for the depth focus model. The port sub-model was the only one to have an affect of decreasing suitability values, and thus can be identified as the primary limiting criterion.

Table 10: Phase II suitability mapping model sensitivity to changes in sub-model inputs analysis results.

Value	Equal Weights	Tidal Focus	Community Focus	Depth Focus	Port Focus
0	41.05%	41.05%	41.05%	41.05%	41.05%
1	0.00%	0.00%	0.00%	0.00%	0.00%
2	0.31%	0.17%	0.70%	0.95%	0.83%
3	3.29%	2.48%	4.54%	4.62%	7.19%
4	15.41%	13.29%	19.85%	19.14%	36.02%
5	32.32%	26.89%	22.03%	11.33%	11.87%
6	5.19%	13.15%	9.51%	17.61%	1.28%
7	2.24%	2.96%	1.40%	4.21%	1.76%
8	0.18%	0.00%	0.93%	1.08%	0.00%
9	0.00%	0.00%	0.00%	0.00%	0.00%
10	0.00%	0.00%	0.00%	0.00%	0.00%

7.2 Candidate community identification

The additional considerations included in Phase II resulted in three communities within the MaPP region becoming economically non-viable due to the community-to-site distance constraint. Bella Bella, Klemtu, and Kulkayu (Hartley Bay) were removed from the analysis leaving the nine communities in Table 11 below.

Table 11: Phase II results and candidate community identification.

Community name	Latitude	Longitude	Distance to closest site	Highest value (≥2 cells)	MaPP Zoning	Site latitude	Site Longitude	Site Umax
Haida Gwaii Sub Region								
Queen Charlotte Village	53° 15' 20.69"N	132° 5' 25.09"W	3.9 km	7	PMZ	53°13'11.043"N	132°5'6.064"W	1.98 m/s
Sandspit	53° 14' 34.54"N	131° 49' 17.13"W	9.4 km	6	PMZ	53°19'22.815"N	131°51'47.761"W	2.12 m/s
Skidegate Landing	53° 14' 53.38"N	132° 0' 41.83"W	5.8 km	7	PMZ	53°13'11.043"N	132°5'6.064"W	1.98 m/s
North Coast Sub Region								
Kitkatla	53° 47' 41.29"N	130° 25' 45.13"W	2.3 km	7	GMZ	53°47'26.524"N	130°23'41.296"W	2.47 m/s
NVI Sub Region								
Tlatlasikwala (Bull Harbour)	50° 55' 22.12"N	127° 56' 21.14"W	4.7 km	7	GMZ	50°53'36.184"N	127°59'1.035"W	2.39 m/s
Kwikwasut'inuxw Haxwa'mis	50° 41' 45.27"N	126° 36' 1.6"W	7.3 km	7	SMZ (tourism and recreation)	50°38'9.852"N	126°33'21.322"W	2.05 m/s
Gilford Island	50° 45' 6.18"N	126° 29' 31.38"W	7.6 km	7	GMZ	50°47'0.954"N	126°35'15.626"W	2.65 m/s
Da'naxda'xw (Tsatsisnukwomi Village)	50° 35' 47.96"N	126° 35' 51.67"W	5.4 km	7	SMZ (tourism and recreation)	50°38'9.852"N	126°33'21.322"W	2.05 m/s
Gwawaenuk (Hopetown)	50° 55' 25.59"N	126° 49' 20.38"W	3.4 km	7	GMZ	50°53'46.917"N	126°50'38.27"W	1.78 m/s

Of these nine communities, only one (Sandspit) had no sites with a suitability value of 7, reducing the number of considered communities to eight. From there, preference is given to sites that fall within a GMZ zoning, screening out four additional communities (Queen Charlotte Village, Skidegate Landing, Kwikwasut'inuxw Haxwa'mis, and Da'naxda'xw). Of the remaining communities, Gilford Island was removed due to the relatively large distance from community to site and Gwawaenuk is removed due to the low peak spring current speed value (1.78 m/s).

This screening left two candidate communities: Kitkatla in the North Coast sub-region and Tlatlasikwala (Bull Harbour) in the North Vancouver Island sub-region. Kitkatla was chosen due to the proximity of the site and larger population, theoretically meaning greater HR capacity. However, when speaking with a representative of Kitkatla it was found that the community had been grid connected for over twenty-five years, pointing to outdated data within the NRCan Remote Community Energy Atlas.

Further attempts at reaching other communities (following the order in which they were screened out) were unsuccessful or stalled (see limitations 10.5.3). The holistic framework of this study required stakeholder engagement to occur in tandem with economic analysis, thus it was decided to seek a community from Table 11 in which both could be accomplished. The interview was eventually undertaken with representative of the Council of the Haida Nation (who was also a previous participant), with Queen Charlotte Village (QCV) and Skidegate Landing being chosen as the candidate communities, despite having a peak spring current speed just below the economic viability threshold, with the Phase II interview taking place over the phone in September 2020 (Nash & Phoenix, 2017; O'Rourke et al., 2010a).

7.3   Candidate community overview

Haida Gwaii (literally translated to "*islands of the people*") is an archipelago on the North Coast of BC with approximately 4,400 inhabitants, half of which are of Haida ancestry (MaPP, 2015b). The are six primary population centres (Masset, Old Masset, Port Clements, QCV, Sandspit, and Skidegate Landing) with electricity service provided by two separate generation-distribution grids as part of BC Hydro's NIA program (BC Hydro, 2012). Masset, Old Masset, and Port Clements are serviced by a diesel generating station in Masset (referred to as the North

Grid); while the South Grid serves QCV, Sandspit, Skidegate, and Tlell via a diesel generating station in Sandspit and a private hydroelectric plant (Moresby Lake Hydroelectric station) located ≈39 km south-west of Sandspit (BC Hydro, 2012). Collectively, the islands use around 10 million litres of diesel annually to produce 65% of the island's electricity (Swiilawiid, n.d.). The South Grid's 1,650 customers get about 75% of their electricity from the Moresby Lake Hydroelectric station (6 MW reservoir-based system owned by an Independent Power Producer-IPP) with the remaining being generated by the Sandspit diesel system (rated prime capacity of 11.25 MW split between 9 units) (BC Hydro, 2020).

Transitioning to renewables has been a long-sought goal on Haida Gwaii, with a community vision over a decade ago to be the first BC community to fully meet electricity needs through renewables and demand management (e.g. energy awareness, efficient appliances) (BC Hydro, 2008). Drivers include improved reliability, an electricity system that allows for economic development/growth, self sufficiency/self reliance, decreasing the Islands carbon footprint, job creation, and the elimination of uncertainty surrounding fluctuating fuel prices (BC Hydro, 2008; Swiilawiid, n.d.). A range of alternatives have been proposed to meet HGs electrical demand including tidal, wind, and solar. Other than the hydroelectric station, RET penetration is currently low, with around 120 kW of small-scale solar capacity on various community buildings within the South Grid and 62 kW within the North Grid (Swiilawiid, n.d.).

Several initiatives exist to transition off diesel on Haida Gwaii, such as the citizen led Swiilawiid Project, and more recently the Haida Gwaii Clean Energy Project (HGCEP) (NRCan, 2019). The Clean Energy for Rural and Remote Communities program will provide funding of $10 million over three years for the HGCEP, with a total project cost of $30 million (NRCan, 2019). Phase 1 includes the expansion of the Moresby Lake station through a dam raise, construction of saddle dams, and commissioning of a previously installed 300 kW turbine (NRCan, 2019). The second phase seeks to develop and deploy a 2 MW solar farm (a joint partnership project between the Skidegate and Old Masset Village Councils called Tll Yahda Energy) on Graham Island (the north island) (Coastal First Nations, 2020; NRCan, 2019). This project will provide enough power for roughly 200 homes and the Tll Yahda Energy partnership will get a stake in the Moresby Lake station as well (Coastal First Nations, 2020). Phase 2 will include the development of other renewables as needed to achieve 100% renewable energy by 2023, as set out in the "People's Clean Energy Declaration for Haida Gwaii" signed by the Haida Nation, Village Councils, Hereditary Leaders, and municipal/regional governments in 2018 (Coastal First Nations, 2020).

QCV and Skidegate Landing have populations of 852 and 837 respectively (Statistics Canada, 2016ab). Despite their closeness, there are several socio-economic discrepancies between the two. Skidegate is a predominantly First Nations community (86%), whereas only 17% of QCV identifies as Aboriginal. While Skidegate's population has grown by 18.1% from 2011-2016, exceeding BCs growth rate of 5.6%, QCV population has decreased by 9.7% (Statistics Canada, 2016ab). Both communities have higher unemployment rates than the provincial average (6.7%) with Skidegate's being much higher (11.7%) than QCV (7.7%) (Statistics Canada, 2016ab). Differences between the two are also apparent when considering median income for full time employment with QCV being $66, 919 (above the provincial average of $53, 940) and Skidegate being $40, 064 (Statistics Canada, 2016ab). This may be in part attributed to QCV having higher education levels with 18% having no diploma, degree or certificate, 23% having a high school education, and 59% having some post secondary education compared to 35%, 27%, and 37% respectively for Skidegate (Statistics Canada, 2016ab). Finally,

Skidegate has a much higher percentage of dwellings in need of major repairs (19%) compared to 14% in QCV and the provincial average of 6.3% (Statistics Canada, 2016ab). This brief overview supports calls within the HG Marine Plan for sustainable economic development opportunities within the communities, while also highlighting socio-economic disparities between predominantly First Nation's and non-indigenous communities.

7.3.1 Candidate tidal site overview

The identified resource is located within Skidegate Inlet (SI), a relatively confined body of water separated from Hecate Strait by a prominent spit (MaPP, 2015b). The area is well known for its high tidal range, abundance of herring, salmon, other fish, and waterfowl (MaPP, 2015b). The Inlet experiences relatively high marine use such as transportation and log/container barge anchorages owing to its proximity to multiple communities (MaPP, 2015b). It is also a transportation corridor between the east and west coasts of Haida Gwaii (MaPP, 2015b). The identified site in Figure 23a and 23b is primarily located within the Xaana Kaahlii SGaagiidaay PMZ, in which MRE is allowed, along with bordering the Laaginda Kaahlii SGaagiidaay aquaculture SMZ and being north of the Gaagun Kun SGaagiidaay aquaculture SMZ, with MRE development being conditionally allowed in both (development may be limited by existing or future shellfish aquaculture operations) (MaPP, 2015b). The Tluu T'aang.nga SGaagiidaay PMZ is also located nearby the site in which MRE is not allowed as the PMZ is meant to protect unique fossils along with eelgrass and Giant Kelp habitat (MaPP, 2015b).

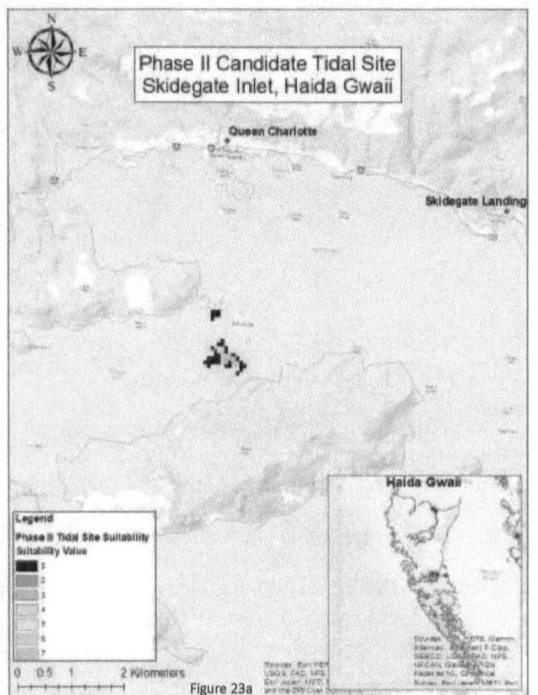

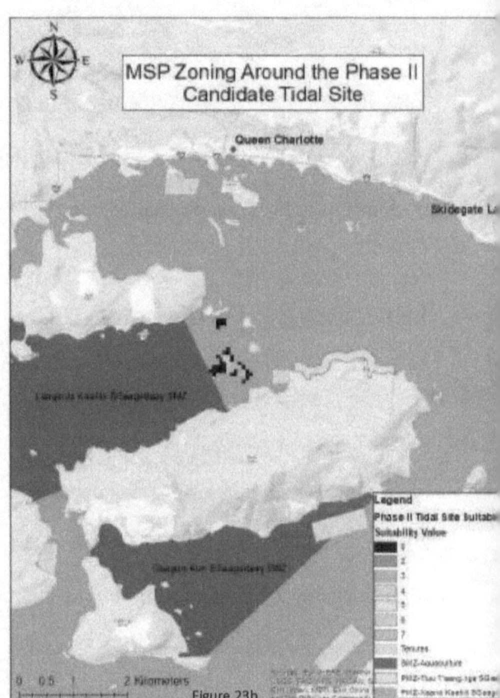

Figure 23: Suitability of potential sites (a) and MSP zoning near QCV and Skidegate Landing (b).

7.4 Phase II interview results

The participant was asked the questions outlined below in Figure 24.

1. Are there any load restrictions imposed by the Sandspit diesel generators in Queen Charlotte/Skidegate?

2. How would you characterize Queen Charlotte/Skidegate's current energy infrastructure? For instance, is it sufficient to meet current demand and how are the diesel generators viewed by the community?

3. What is the current cost of energy production in Queen Charlotte/Skidegate (i.e. $/kWh or $/MWh?)

4. How do you foresee community energy needs changing over the next twenty years? (e.g. will the community require more energy due to increasing population, infrastructure upgrades, and economic development goals?)

5. What diesel alternatives have been considered in the past? Why were they successful or why were they not successful?

6. What are the characteristics of a potential energy project that the Council of the Haida Nation looks for? (e.g. ownership over assets, public vs private partnership, community employment, etc.)

7. How would you characterize human resource capacity in Queen Charlotte/Skidegate (or Haida Gwaii more broadly) to champion alternative energy options, apply for funding, participate in the development of alternative energy sources etc.?

8. Based on the maps provided, and your knowledge of the area, is the identified tidal site a suitable location for development? That is, does it minimize infringement on other social, economic, and environmental uses? Please answer on a scale of 1-5 (1 strongly disagree, 2 disagree, 3 unsure, 4 agree, and 5 strongly agree):

9. Regardless of perceived suitability, please comment on the location of the site and what community uses and values a potential device would come into conflict with if deployed there?

10. What data sources could be added to make this map more representative of the communities' interests and uses in the area?

Figure 24: Phase II candidate community interview questions.

The supply-demand balance on the islands seems appropriate to the participant and they were unaware if there were any grid-imposed restrictions on the generators. Likewise, from their perspective as a customer there had not been issues or anything that they had become aware of while engaging in energy discourse. They did note the discrepancy in how diesel generators are viewed between the South Grid and the North Grid, with the North Grid having a greater visual impact as the diesel station is visible from a prominent roadway. Overall, they noted that the diesel generators are generally viewed negatively, driven by community reliance on them and associated climate impacts along with a strong drive for the Islands to become self sufficient.

The participant was not able to provide cost data, and thus BC Hydro was contacted (see section 8.1 for more detail).

In terms of future energy needs, they hoped that demand would remain similar to current levels or decrease. The participant explained that this is potentially realistic given community groups (i.e. the Skidegate Band Council) becoming more energy self sufficient, increasingly energy efficient technologies becoming available, and Islanders being ever more cognisant of energy related issues. Finally, they noted that the size of industry on HG is relatively small, along with a general trend towards tourism and less energy intensive industries, meaning energy would likely stay the same or decrease into the future. This was discussed with the caveat that there is always the possibility for a business that is very desirable to the community being proposed and increasing demand.

Regarding alternative energy sources, they described the success of community solar power projects, specifically on the community hall and the Haida Heritage Centre, along with the installation of energy efficient technologies such as heat pumps. These successes ranged from the projects being talking points to develop awareness and energy education along with positive economic benefits in terms of job creation and reduction of electrical bills.

In terms of tidal, they noted that the local project had not been successful thus far. They attributed this to a divide between project planners and some of the process that are in place at the CHN. They emphasized that the lack of progress is not the onus of one organization or the other, but a lack of capacity for both organizations to find a suitable (primarily avoiding cultural and environmental impacts) site to develop which has been the sticking point delaying the project.

Characteristics the CHN would look for in a tidal energy project was primarily focused on compliance with existing MSP and zoning. Second to that is the need for Haida first and local community employment.

HR capacity was viewed as being good, especially given the population size and remoteness of HG. While it was stated that there aren't many engineers on island, there are a few along with people who might have the necessary skills for installing devices and sourcing materials or would have contacts who could. The Skidegate Band Council was described as championing renewable projects along with representatives/employees attending conferences and making connections with the renewable energy community. Capacity was viewed as "good" when considering the ability to apply for funding and championing RET projects, and "great" in terms of willingness to participate in projects.

When it came to assessing the suitability of the identified site, they noted that they could not answer that kind of question on their own, citing the need to include special experts for different uses from within their team. They did mention that referring to the existing zonation through the HG marine plan is a good place to start, although they are in the process of undertaking more detailed zoning and planning in SI. They also noted that the area is frequented by boating activity (fishing, navigation, recreation) along with aquaculture tenures so there would be some conflict, but the site might be suitable if the conflict could be managed. Other data sources that would make the maps more indicative of marine uses include: the Haida Gwaii Marine Traditional Use study, along with an internal interactive map, a stream map indicating important salmon habitat, and shellfish aquaculture carrying capacity studies.

Chapter VIII: Phase III methods

Phase III of the study involved three LCOE analyses: an existing diesel generation LCOE, an externality included LCOE for existing diesel generation, and a tidal LCOE. This allowed for direct comparisons of existing infrastructure to that of a potential TST development.

8.1 Diesel LCOE analysis

This research benefited from the disclosure of electricity load data from BC Hydro for the South Grid. This included annual generation in kWh for both the Sandspit diesel station and Moresby Lake station over 2014-2019 along with information on generator make, model, and prime power ratings. Annual fuel costs were calculated according to Equation 8.1:

$$F_{OPEX} = \frac{E}{F_r} \times F_c \qquad 8.1$$

Where E is the annual electricity production, F_r is the average fuel consumption rate (3.5 kWh/L taken from the Hatch report and supported by discussions with BC Hydro), and F_c is the cost of fuel ($1.3/L) (Hatch Energy Ltd., 2008).

Through discussions with BC Hydro annual non-fuel OPEX was said to be in the range of 46% of total OPEX. The annual OPEX was calculated based on the annual fuel cost and multiplied by 0.46 to obtain the non-fuel OPEX. The LCOE was calculated in Excel using Equation 2.1 and the inputs in Table 12 below.

Table 12: Sandspit diesel generator LCOE inputs.

Input	Metric
Average annual electricity production	5,841,673 kWh
Annual Fuel costs	$2,169,764 Along with high fuel price ($1.5/L) and low fuel price ($1.1/L) scenarios
Annual OPEX	$2,074,056/year
CAPEX	Not included in LCOE assessments for diesel generators by BC Hydro
Project lifetime	20 years
Average inflation rate	1.88%
Discount rates	4% and 10%

8.2   Externality included LCOE analysis

As described earlier, the inclusion of externalities within LCOE analyses provides a more encompassing picture of costs associated with the energy system. CO_2 emissions, more specifically CO_2 equivalent (CO_2e) emissions (quantifying the radiative forcing of all GHGs emitted by a source to a common scale), represents arguably the greatest environmental, economic, and social cost associated with diesel generated electricity. Accounting for the cost of each additional tonne of carbon is complex owing to the vast number of uncertainties associated with future climate change impacts resulting from differing emissions trajectories (i.e. relative concentration pathways in Intergovernmental Panel on Climate Change reports). Table 13 below provides an overview of international, national, sub-national, and academic studies on the cost of carbon.

Table 13: Values for carbon per tonne. Taken from academic studies and government carbon tax values.

Author(s) or Organization	Social cost of carbon or carbon tax value (expressed in 2020 \$CAD/tonne if values were calculated before 2017)
Intergovernmental Panel on Climate Change, 2007	\$57
Stern, 2007	\$112
US Interagency Working Group on Social Cost of Carbon, 2010	\$62
Balbus et al., 2014	\$48-219 Positive benefits per tonne mitigated by 2020.
Government of Canada, 2017	\$30 Rising by \$10 to \$50 in 2022.
Ricke et al., 2018	\$540
Cali & Lontzek, 2019	\$81-137
van der Ploeg & de Zeeuw, 2019	\$20-94
Government of British Columbia, 2020	\$40 Was scheduled to rise to \$45 in April 2020. As a result of the COVID-19 pandemic it remained at \$40. Rising to \$50 by July 2021.

This study utilized a carbon tax of \$50/tonne, in keeping with British Columbian and Canadian 2021/2022. While this is on the lower end of valuations, it can be taken as a willingness to pay benchmark price as it is grounded in a legal regulatory framework. LCOEs were also calculated for carbon taxes of \$75 and \$100, respectively. In actuality, the 'true costs' of carbon are likely much higher, with carbon taxes being simple constructs applied by jurisdictions to try and encapsulate some of these costs.

The cost of annual CO_2e emissions was calculated by multiplying annual fuel consumption by the ratio of kilograms of CO_2e emitted per liter of diesel fuel combusted, 2.705 $kgCO_2e/L$, multiplied by the price of the carbon tax (MECCS, 2019). This was then included in non-fuel OPEX costs within the diesel LCOE calculator.

8.3   Tidal LCOE analysis

NRCan's *Tidal Project Cost Estimation* tool was used to calculate the LCOE for the TST at the SI site. Project and turbine specifications are defined by the user and input into the tool along with resource data for the site to yield an LCOE value. The input categories are: Project specifications (Table 14), Turbine specifications (Table 15), and Site specifications (Table 16) (Green Kappa Consulting and Technology Inc., 2018). For more details on specifications rationale see Appendix L.

8.3.1 Project specifications

Table 14: Skidegate Inlet project specifications.

Specification	Metric
Project lifetime	20 years
Number of turbine rows	1
Number of turbines	4
Project start year	2023
Average inflation rate	1.88%
Grid connected	Yes
Discount rate	Social discount rate of 4%, technology discount rate of 10%
CAD to USD conversion rate	1CAD = 0.76USD

8.3.2 Turbine specifications

A TST needed to be chosen that matched the site resource, for this, the Schottel STG 54 was chosen. It is flexible in terms of its application (can be bottom mounted or floating), modular in that multiple turbines could be installed on a single platform (such as Sustainable Marine Energies PLAT-I deployed at FORCE) and has the lowest cut in speed and rated speed of the devices surveyed (refer to Appendix D for more information). Four turbines are deployed decrease $/kW installed, the same set up as PLAT-I (Figure 25 below).

Figure 25: PLAT-I being worked on at FORCE. Taken from https://www.sustainablemarine.com/plat-i.

Table 15: Skidegate Inlet TST technical specifications.

Specification	Metric
Type of turbine	Horizontal axis
Horizontal axis turbine diameter	5 m
Cut-in speed	0.7 m/s
Rated speed	2.6 m/s
Cut-out speed	4.6 m/s
Maintenance frequency (every # years)	3
Turbine hub position	4.5 m
Turbine power coefficient	0.4
Is turbine shrouded?	No
Wake effect coefficient	0.85
Deployment type	Floating
Foundation type	Floating two points
Yaw mechanism?	Yes

8.3.3 Site specifications

Table 16: Skidegate Inlet site specifications.

Specification	Metric
Average depth	17.2 m
Water density	1027 kg/m^3
Seabed Type	Shallow Sediment Layer
Distance to shore	3.9 km
Distance site to port	167 km
Distance site to grid	1.00 km
Distance site to harbour (for moorings vessel)	15 km
Distance site to harbour (for cable laying vessel)	15 km

Chapter IX: Phase III results

9.1   Existing diesel infrastructure LCOE

Table 17: Sandspit Diesel Generating station LCOE for existing infrastructure, varied discount rates, and high and low fuel costs.

Scenario	LCOE
Existing setup at a 4% discount rate	$0.63/kWh
Existing setup at a 10% discount rate	$0.76/kWh
High fuel cost of $1.5/L at a 4% discount	$0.72/kWh
Low fuel cost of $1.1/L at a 4% discount	$0.53/kWh

The current price of electricity sold to ratepayers ($0.1121/kWh for the first 1500kWh/month, $0.1925/kWh thereafter) indicates BC Hydro is operating at a loss of at least $0.43/kWh of energy produced and delivered to the HG South Grid at current fuel prices.

9.2   CO$_2$ equivalent emissions externality LCOE

Table 18: Externality included LCOEs

Scenario	LCOE
$50/tonne carbon tax at a 4% discount rate	$0.65/kWh
$50/tonne carbon tax at a 10% discount rate	$0.79/kWh
$50/tonne carbon tax at a 4% (10%) discount rate and $1.5/L fuel price	$0.75kWh ($0.91/kWh)
$75/tonne carbon tax at a 4% discount rate	$0.66/kWh
$100/tonne carbon tax at a 4% discount rate	$0.68/kWh

The results displayed in Table 17 & 18 indicate that a TST device would require an LCOE below $0.63/kWh ($0.65/kWh including carbon externalities), to a maximum allowable LCOE of $0.91/kWh based on future carbon tax rates and a high discount rate applied.

9.3   Skidegate Inlet tidal site LCOE

Table 19 highlights the complete economic non-viability of the SI tidal site, with a LCOE

of \$24.04/kWh, over thirty-eight times the current diesel LCOE and twenty-six times the

maximum "allowable" LCOE for South Grid energy production.

Table 19: Results of the Skidegate Inlet TST LCOE analysis.

Metric	Value
Platform rated capacity	284 kW
Annual Energy production	40,100 kWh
CAPEX	\$4,367,790
Annual OPEX	\$385,275
CAPEX/kW	\$15,405
Capacity factor	1.62%
LCOE	**\$24.04/kWh (\$2,404/MWh)**

This can be primarily attributed to the distribution of current flow at the site, with a peak

frequency distribution around 0.81 m/s while also having a peak site velocity (calculated within

the tool at 1.79 m/s) far below the TSTs rated current speed (as illustrated in Figure 26), resulting

in the device being characterized by a capacity factor of 1.62%. Figure 27 breaks down CAPEX,

with nearly a third of capital costs attributed to cabling. When considering a platform with only a

single turbine, cabling represented over half of CAPEX. Based on Figure 27, the total cost of the

turbines is \$1.048M, approximately \$262,000/turbine installed (\$3,690/kW of rated capacity).

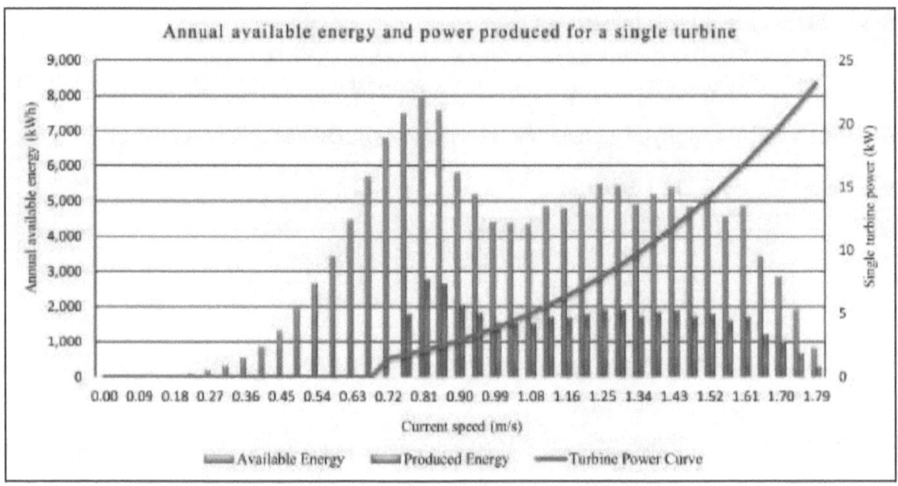

Figure 26: Annual available energy and power produced by a single turbine at the Skidegate Inlet TST site. Taken from NRCan's Tidal Project Cost Estimation tool.

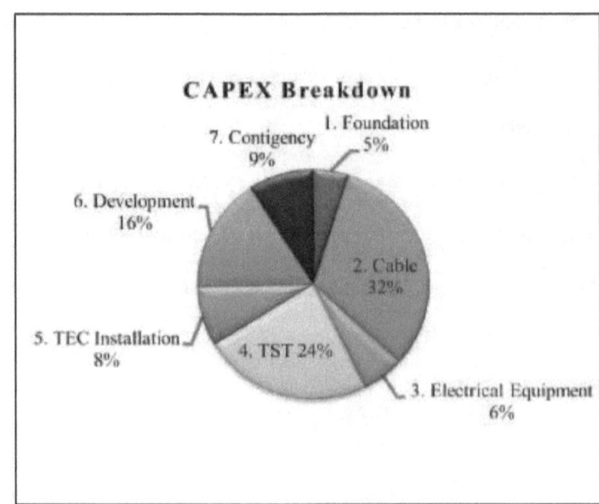

Figure 27: Skidegate Inlet TST capital expenditures breakdown by cost category. Taken from NRCan's Tidal Project Cost Estimation tool.

Chapter X: Discussion

10.1 Electricity in remote coastal diesel reliant First Nations communities

The Phase I and II interview findings supported much of the existing literature in regard to drivers for renewable energy adoption in RCDRC, especially the importance of energy self-sufficiency as a driver for community electrification. There were some discrepancies between findings and existing literature in terms of drivers. These included frequently cited diesel externalities in the literature such as health impacts, associated emissions, spill risk, and the reliability of diesel generator operation, which were not found by this research to be among the most prominent drivers. The Phase II interview did emphasize climate change as a major driver for HG.

The importance of reducing energy costs was viewed as being more important by FLNRO participants than by FN. This focus can be attributed to ownership of generating assets in communities served under the NIAs being the responsibility of BC Hydro, and in kind the provincial government, and thus they have an incentive to reduce costs as much as possible.

Tidal energy appears to be favorably viewed from the perspective of those involved in higher levels of planning within the MaPP. Many respondents noted their disappointment with the lack of tidal development in the MaPP area given its potential during initial stages of planning. Others stated the need for further research on tidal energy and expressed their organizations interest in the findings of such work. Such results point towards considerable interest for tidal energy within the MaPP region, both for community electrification and for powering remote economic operations.

A suitable TST system would need to be reliable, have a high-capacity factor, and would require a back up power system (or sufficient storage). Participants also described the need for the technology to be capable of being fixed/troubleshot by people residing within the community. Other respondents noted the need for community control over generating assets alongside job provision and learning opportunities for the community. Furthermore, communities are looking for collaborative reciprocal partnerships with government and industry. Lastly, a tidal energy system would have to be compatible with existing marine uses.

10.1.1 Barriers to tidal energy electrification and pathways forward

While tidal appears to be a promising RET, there are several barriers to its implementation in RCDRC. Firstly, there is a huge lack of information and accessibility to that information for communities to understand or assess whether TST technology is feasible and a good fit. For instance, FN in comparison to FLNRO respondents were on average not familiar with how the technology works, the spatial requirements, environmental impacts, regulatory requirements, and especially the cost of energy. This may be partially attributed to a lack of community access to available data and to resources such as academic literature and technical reports, with FLNRO employees likely having greater access to these resources along with the HR capacity to review them. Perhaps not surprisingly the two highest average responses were from HG based participants, owing in part as participant 4 noted, the CHN and FLNROs close working relationship with a HG based tidal company. The lack of familiarity and understanding of TSTs is a huge barrier to development in the MaPP region. For instance, one participant emphasized that their concerns surrounding the environmental impact of TST devices might just be misconceptions that could be alleviated if communities were privy to up-to-date research.

FN representatives identified information to assess RET options as the second greatest perceived barrier to community electrification. Not only on the economic/technical feasibility side, but as participant 6 highlighted, the importance of understanding social and cultural views towards RETs. Multiple participants identified the need for sufficient HR capacity (or leveraging of external resources) to undertake such assessments and to understand how existing renewable assessments apply to a community. This should go beyond TSTs, as tidal must not only compete with existing diesel, but with a suite of RETs as well if distinct communities are to utilize the most suitable (in terms of community values), reliable, and cost-efficient electricity source.

Funding was identified as the greatest perceived barrier. Not only in terms of the capital needed to develop projects, but also funding to enhance a community's capacity to navigate project assessment/development. While there is substantial funding available for electrification projects, both from the federal and provincial governments, participants noted the difficulties in navigating funding opportunities especially across different government agencies (Government of Canada, 2019; Province of British Columbia, 2018). Furthermore, HR capacity again is identified as a barrier in this respect, as communities require the capacity to pursue these funding opportunities in the first place. HR for OPEX may also be a barrier to consider, as despite the promise of TST technologies they are still relatively novel, being associated with more logistically challenging operation and maintenance requirements than land-based generators. A recent report on the operations of Meygen Phase 1A supports this notion, even for large companies with access to significant resources (Black & Veatch, 2020). While HR appears to be a significant barrier especially in preliminary RET assessments for RCDRC, this was not the case for all, with the Phase II interview indicating appropriate HR capacity on HG for funding applications and championing of projects.

10.2   Integrating tidal energy into MaPP region

The MaPP region boasts significant tidal resources, with approximately 89.8 km^2 of technically feasible and constraint free area. These sites tend to be concentrated in narrow passages and constrained water ways between islands (refer to section 5.3).The NVI sub-region had the greatest areal coverage by far ($\approx$62.7 km^2), followed by HG ($\approx$20.1 km^2), the CC ($\approx$3.8 km^2), and the NC sub-region ($\approx$3.1 km^2). When considering tidal energy suitability specifically for RCDRC, the total suitable area decreases to only 25% ($\approx$21.9 km^2) of the total feasible area.

10.2.1 Incorporating marine uses into tidal energy suitability analyses

It was critical from multiple perspectives to include uses of the marine space and ensure that tidal development occurs in harmony with these uses. In terms of the perceived potential for conflict with tidal energy, commercial fisheries were given the greatest weighting out of economic uses (see Appendix G). This is potentially a result of the fisheries using wide swathes of the marine space, while also being a dynamic use (i.e. fleets or vessels may transit numerous and varied areas whereas larger scale shipping or transportation is commonly constrained to specific transects). The second greatest perceived conflict was government and research vessel traffic, although it had the greatest range in responses which indicates some uncertainty in how the use is viewed. Uses such as merchant vessel and tanker vessel traffic appear unlikely to ovelap with tidal resource locations, as their transects are predominantly in more open waters such as Hecate Strait (separating HG and the mainland).

There was a greater diversity in the available economic uses datasets compared to social uses assessed in this study (six layers versus three), with the economic layers being arguably more comprehensive in terms of representing categorical uses. For instance, small vessel traffic was a notably absent use in terms of data availability, while also being a use that likely constitutes a substantial amount of marine activity in areas of the MaPP region. The most overarching of these layers, the tourism and recreation feature count, covers a wide range of uses and was clearly viewed as being the most important, likely due to it acting as an encompassing proxy for the most uses. Furthermore, the decision to allocate a value of 8 to 'no data' areas of the sport fishing and recreation/tourism feature counts in order to ensure coverage of sub-models across sites had the effect of increasing suitability for the Social Uses sub-model. This contributed to greater overall suitability values for models weighted more heavily to social uses.

It was evident from the perspective of high-level MSP that environmental impacts are a primary concern regarding tidal development and device siting. All three FN respondents ranked Marine Conservation Value as the most important sub-model input after the tidal resource, while FLNRO participants ranked it as 3rd and 4th respectively. The second greatest consideration appears to be social uses, weighted more heavily than economic uses in the AHP results while also being consistently identified in participant responses. Other areas of concern included navigability of the marine space and safety, although these were viewed as being of minor concern when considering TST integration.

10.2.2 Future planning for tidal energy in the MaPP region

Four of the five tidal energy SMZs within the MaPP region had a feasible resource identified in this study, with one (Galloway Rapids located south of Prince Rupert) being characterized as technically unfeasible. Additionally, there were numerous sites identified within favorable zoning (i.e. alternative energy is allowed or conditionally allowed) that might merit delineation as tidal energy SMZs if further scrutinized. The results of this study can contribute to future iterations of the MaPP sub-regional plans. Future work could also examine the proximity of suitable tidal resources to economic tenures, thus supporting the delineation of additional alternative energy SMZs to power remote operations. This would greatly improve the spatial delineation potential TST sites in the MaPP region while providing communities, industry, and government with a better understanding of where opportunities might be.

One of the greatest issues with this study, especially in Phase I, was the lack of cultural dataset accessibility and subsequently an absence of cultural uses being represented within models. This was highlighted by multiple participants as a shortcoming of the Social Uses sub-model, while the importance of cultural uses in determining suitable locations for tidal development was continually emphasized. The lack of integrated cultural data was attributed in part to the difficulties as a single graduate student researcher to foster and build timely relationships with a range of First Nations and obtain potentially culturally sensitive data. Likewise, available datasets such as cultural emphasis SMZs do not provide an indication of relative use beyond a Boolean analysis and therefore it was decided instead to identify cultural data as being a prominent data deficit. Integrating cultural and supplementary data into assessments may be more effective and easier to implement at community scales, as illustrated by the additional resources cited by the participant in the Phase II interview.

More broadly, future tidal energy suitability mapping and MSP as a whole in the MaPP region and along the BC coast would benefit greatly from updated data. There is significant need for contemporary datasets which provide a greater understanding of the marine space as the utility of many of the existing datasets is diminishing as they have already exceeded their useful timeframes. This is discussed in more detail within the study limitations in section 10.5.1.

10.3   Candidate community case study

10.3.1 MSP suitability of the identified site

The SI site appears to have suitable MSP for tidal development based on the Phase II interview and analysis of the area referring to the Haida Gwaii Marine Plan. Although the site is within a PMZ, it is zoned as a low protection area in which alternative energy development is acceptable. The site is considered high value in terms of commercial fisheries for herring spawn on kelp and salmon, while transportation activities and temporary vessel anchorages (primarily by log and container vessels) have high economic value in the surrounding area (MaPP, 2015b). The nearby Laaginda Kaahlii shellfish aquaculture SMZ may present both conflicts (e.g. impedance of associated vessels) and synergies (e.g. supplying power to operations). A complete assessment of suitability remains a subject in need of more in-depth analysis in partnership with the CHN (as demonstrated by the Phase II interview) to determine whether tidal development would be allowed, but given the ability to reduce conflicts with human uses the site appears to be suitable.

Conversely, the proposed tidal site at Justkatla Narrows on Graham Island has encountered issues related to perceived cultural and environmental impacts. Justkatla Narrows is zoned as medium protection with a focus on the conservation of kelp beds and eelgrass meadows that provide critical habitat for a range of species along with being an important area for Haida traditional uses and values (MaPP, 2015b). Alternative energy development is conditionally allowed in this area. Those conditions being: 1. Development should avoid disturbing sensitive or critical features of habitat and 2. The activity should be compatible with Haida cultural use. Given the results of the Phase II interview, these conditions are clearly a persistent barrier to development, and it appears that the SI site would be more suitable from a zoning and MSP perspective than Justkatla Narrows.

10.3.2 Economic viability of the Skidegate Inlet site

Despite high cost of diesel electricity generation for the South grid ($0.63/kWh) and the apparent MSP suitability of the identified site, the SI site is economically unviable. Even at BCs and Canada's 2022 carbon tax value the difference in LCOE is substantial, at over $23 more per kWh. Furthermore, the TST device would do little to displace overall diesel use, offsetting 40,100 kWh of electricity and abating around forty tonnes of CO_2e annually ($\approx 0.009\%$ of annual emissions from the Sandspit diesel station).

The two primary factors driving these results are:

1. Poor resource at the identified site resulting in the TST device being characterized by a very low-capacity factor of 1.62%.
2. High cabling costs due to site distance from shore/the communities and cabling installation cost assumptions.

This suggests that while the rationale for omitting QCV/Skidegate based on zoning was not warranted (as the PMZ in which the SI site is located conditionally allows MRE), the rationale of 'ranking' communities as being most suitable based on highest peak spring current speeds and proximity to communities was. These results are in keeping with other findings in the literature, suggesting technical viability (and certainly economic viability) requires peak spring current speeds greater than 2 m/s (see Peak Spring Current Speed constraint in Appendix B). It also must be noted that CAPEX was not included in the diesel LCOE calculations as BC Hydro has a fleet of back up generators/perform overhaul on devices. The inclusion of CAPEX for diesel systems would raise LCOE for the Sandspit station and any other RCDRC serviced by BC Hydro.

Optimally in situ field measurements would also be undertaken to verify modeled current speeds, this represents a significant area of future research as there were issues identified with the tidal resource model, especially around HG, detailed in section 10.6.1. Future iterations of this framework and other studies investigating economically feasible sites should increase the minimum viable peak spring current speed from 1.5 m/s to 2 m/s at the very least or consider the mean peak spring current speed as the screening metric to be used.

While the SI site is not viable, HGs price of electricity and higher prices found in other RCDRC such as Hartley Bay ($0.74/kWh), are still within the range of cited LCOEs for TSTs given a suitable resource ($0.49-55/kWh). Based on the distance and resource constraints identified in the economic analysis, it would appear Bull Harbour would be the most suitable RCDRC for electrification via tidal out of those listed in Table 11, with Kitkatla being a potentially suitable grid connected community. A greater understanding of cabling costs may prevent high resource potential communities such as Gilford Island from being screened out. More work is required to quantify more accurate cabling costs by integrating cost-distance pathway considerations to determine the lowest cost route beyond Euclidean distance. For example, despite Gilford Island having a site to community distance over 7 km, only 2.2 km of that distance would require submarine cables.

10.3.3 TSTs and Haida Gwaii's energy future

More work needs to be done to quantify the tidal resource in Masset inlet, and more importantly examine whether a device would be permissible given the persistent issues with site identification. There are also other sites which this book did not touch upon but have been identified in other analyses, such as Cook Point in Masset Inlet (Hatch Energy Ltd., 2008). Future work would require either in-situ site data, or the refinement of the NRCan Canadian West Coast Tidal Resource Assessment model to examine site resource potential.

Future considerations for tidal development on HG and for any community powered by BC Hydro will be further constrained as BC Hydro does not currently have a standing offer program to purchase power from an independent power producer or community (to power themselves), does not provide incentives for communities to develop renewables (e.g. will not fund a project that offsets diesel use as they require total system redundancy and will not provide a feed in tariff similar to the one present at FORCE in Nova Scotia), and finally the project developers would solely be responsible for all cabling costs (demonstrated to be a significant component of project costs). BC Hydro's present stance poses a significant barrier to community electrification and self sufficiency.

The current trajectory of the HGCEP and Tll Yahda Energy's plans indicate that tidal energy assessments in Masset Inlet were needed 'yesterday'. Given that the capacity factor for solar is in the range of 7-10% on Graham Island and production is heavily skewed to summer months (with peak demand occurring in the winter) there is still need for baseload renewable capacity on the island if a complete departure from diesel is to be achieved (NRCan, 2020b). The proposed solar farm would produce around 1,226-1,752 MWh/year; thus, a potential tidal turbine (or cumulative turbines) would require a rated capacity of 1 MW at a capacity factor of at least 20% to produce the same amount of power, which would certainly be feasible given sufficient natural resource.

10.4   Levelized Cost of Energy shortcomings

Cost has been identified in the literature as a substantial barrier to TST development, a finding also supported by this book. While LCOE analyses do provide the most suitable comparison of dissimilar energy technologies of differing rated capacities, they fail in several aspects while also potentially presenting an inflated cost for tidal relative to diesel given the assumptions and data used in this study. These issues range from assumptions within LCOE analyses themselves, the inherent economic focus of LCOE analyses, and most importantly shortcomings within the NRCan Tidal Project Cost Estimation tool.

10.4.1 Levelized cost of energy analysis shortcomings

For one, traditional LCOE analyses assume that the value of electricity is fixed across the lifetime of the system and they fail to incorporate the parameters of agreements such as PPAs (which are especially relevant in the BC jurisdiction) (Bruck et al., 2018; Dalton et al., 2015). As a result, LCOE analyses rely on a set of static parameters which are not indicative of dynamic real-world scenarios. Another example is the recently disclosed information that the federal carbon tax will now be rising to $170/tonne by 2030, which would of course have implications on the externality included LCOE (Tasker, 2020).

While the use of a high discount rate for TSTs and a low discount rate for diesel generators reflects the existing view of these technologies, this does not incorporate social costs/benefits of developing renewables and the future impacts of using diesel fuels for electricity. In this sense it could be argued that those technologies which will have a greater future impact on climate should be given a more significant discount rate compared to those that do not. The way the study was structured fits with existing financing paradigms for assessment but is not necessarily indicative of how energy production should be assessed in the face of climate change. As noted earlier, discount rates have a significant impact on the outcomes of LCOE analyses and there are suggestions that rates used should be geographically distinct based on the perceived risk at a site and also local socio-economic values/wants (Dalton et al., 2015). In the case of HG, the community drive for self sufficiency and to produce emissions free power support the notion that tidal should be assessed at a lower social discount rate (e.g. around 4%), while diesel should be assessed at a higher rate.

Furthermore, LCOE analyses are inherently financially focused on project costs, and while this study sought to being including externalities within cost estimations it barely scratched the surface of what could (and should) be considered for holistic assessments of energy systems. In addition, LCOE analyses fail to incorporate the total economic value and non-market costs and benefits of investing in tidal or other MRE technologies, which is a major barrier to industry development (Polis et al., 2017). Quantifying the impacts of job creation, research benefits, and more, all of which might greatly enhance the suitability of tidal energy and reduce 'apparent' costs, especially when evaluating integration from a public perspective.

10.4.2 Natural Resources Canada Tidal Project Cost Estimation Tool

Use of the NRCan Tidal Project Cost Estimation Tool was associated with several assumptions and shortcomings that contributed to the large LCOE found at the SI site. For one, assumptions such as the inability to specify a turbine exactly (with difference in the modeled attributes of the STG turbine compared to its actual listed attributes) hamper the analysis. The tool also does not allow for a rated capacity of a device to be specified, as it is generated based on device cut in, cut out, and rate speed along with the site resource. While the tool is rightfully focused on the most advanced tidal technologies, there are also emerging technologies such as tidal kites which are more suitable for low current speeds that may be much more viable at a site such as SI (IRENA, 2020).

The project cost at the SI site can be reduced by 59% to $9.89/kWh by shrouding the turbine and using a social discount rate of 4%. Such results demonstrate that the economic viability of the SI site can be manipulated quite easily in the favour of tidal, despite the poor resource at the site while also demonstrating the vital importance of taking this analysis as a starting point for economic valuations of tidal in BC. Given a suitable resource in which the TST device could achieve a capacity factor of 35% (which is at the conservative end of the cited 30-54% range), the TST LCOE at the SI site would be reduced to $0.99/kWh and $0.67/kWh at 10% and 4% discount rates, respectively. This illustrates how a more suitable resource would significantly reduce LCOE.

Most importantly, the NRCan tool uses TST device capital cost data from 2012 (Green Kappa Consulting and Technology Inc., 2018). As capital costs represent 70% of overall LCOE, updated device data (especially if it is obtained working in partnership with developers) would greatly reduce LCOE results and enhance the potential viability of tidal energy at the SI site and the BC coast more broadly (Vazquez & Iglesias, 2016b). Furthermore, the European Commission's Joint Research Centre found a 40% reduction in LCOE in the last three years alone, meaning these costs are further outdated (Villate et al., 2020).

10.5   Developing and supporting the tidal energy industry in BC

As this study has investigated tidal suitability for RCDRC for nearly two-thirds of the BC coast it is necessary to examine and provide recommendations for the industry in BC in order to inform next steps and avoid previous pitfalls.

Despite emerging tidal companies and research undertaken by groups such as the Pacific Regional Institute for Marine Energy Discovery in recent years, the long-term trend of the industry in BC has been static. Given the long-time frame since the significant resource was identified in BC (18+ years), and the concurrent takeoff of tidal energy on Canada's east coast, there has been a serious lack of progress on the BC coast. This lack of mobilization was never focused on in the interviews, though previous work has identified government support and buy in as a strong barrier to the development of the industry (Richardson, 2018). Additional factors identified include a more general incomplete understanding of TSTs, from environmental impacts to costs. However, the province can still take advantage of resource-demand synergies on the west coast. There are a plethora of research and development opportunities that could be realized with sufficient interest from local groups/communities and government support. For instance, the permitting and development of a research/test site for demonstration of technologies on the BC Coast near tidal resources and with sufficient industrial/HR capacity (e.g. Campbell River) could bring substantial benefits through industry development and exportation of research.

10.5.1 Opportunities for community self-sufficiency through tidal energy

As identified within this study and others, self sufficiency and control over community electricity is a central driving tenet for the electrification of diesel reliant FN communities. For those communities not identified as being candidates for TST electrification (and those that have), there still may be alternative opportunities for tidal development within their territories. First and foremost are opportunities for powering remote operations that are in proximity to resources such as aquaculture and logging operations. These would provide collaborative synergies between FN and industry while providing economic opportunities to nearby communities.

Another potential opportunity is that of tidal to power ocean monitoring installations, as outlined in the *Powering the Blue Economy* report (LiVecchi, et al., 2019). This would provide the opportunity to protect some of the richest marine ecosystems on earth, facilitate the collection of oceanographic data as climate change impacts continue to manifest, while also pioneering FN-government-academia-and industry partnerships to monitor ocean health using novel technologies. Opportunities and funding already exist, such as Canada's Ocean Supercluster recently allocating $29 million to the Ocean Aware project, intended to develop and commercialize environmental monitoring solutions to support profitable and sustainable ocean uses (COS, 2020). This also presents opportunities for collaboration and co-management with existing FN programs, such as the Coastal First Nations Coastal Guardian Watchmen.

Tidal energy development could also support the growth of new industries such as hydrogen electrolysis from water for fuel creation (LiVecchi et al., 2019). This could further foster self sufficiency on HG as the hydrogen produced could be used as energy storage in fuel cells or as a fuel source for marine and terrestrial vehicles. Such initiatives are already being trialed at EMEC in the outer Orkney Islands off the northeast coast of Scotland (LiVecchi et al., 2019). Another emerging demand for power is the development of direct carbon capture endeavours. Installing TST supply capacity beyond community demand would allow for the opportunity to sequester atmospheric carbon (again taking the carbon tax as a regulatory rate for abatement) or the captured carbon could be used to make synthetic fuel for island use, both of which would create new economic opportunities for HG and potentially for other communities (although larger communities are much more likely to be able to support such endeavours. With increased demand from emerging industries, larger economies of scale could be reached, in turn reducing the cost of electricity and ensuring any excess energy can be allocated to useful applications. Other use examples include sea water mineral mining, powering underwater data centers, desalination, and more (LiVecchi et al., 2019).

In summary, tidal energy has the potential to not only provide power to remote communities in the MaPP region but also provides a plethora of economic opportunities for coastal communities, a significant goal of MaPP (MaPP, 2016). To progress towards such an innovative future requires more thorough analysis of potential synergies and most importantly the testing and development of devices in the MaPP region.

10.6 Limitations

10.6.1 Data, processing, and methods

The results of GIS-MCDA analyses are heavily dependant upon the availability and suitability of data to represent criteria and constraints. Insufficient or unsuitable data can produce results which are not representative of the real world, greatly reducing the effectiveness of the fundamental goal of utilizing such methods in decision making.

NRCan Canadian West Coast Tidal Resource Assessment model

The tidal resource dataset used in this study had several shortcomings. Firstly, there were issues with the model operating in some of the narrower channels, especially around HG, where a different bathymetry source was used compared to the rest of the coast. This may be one of the reasons for the poor resource identified around HG, and why several areas of significant tidal flow that were identified within the Hatch Energy Ltd. report did not appear in this study. Validation of this dataset with in-situ measurements would provide greater confidence in the model's ability to identify resource locations, along with allowing for refinement in future iterations of the model.

Uses of the marine space

Other data limitations can be pointed at the outdated marine use datasets taken from the BCMCA. While the BCMCA' multi-year effort (2006-2013) collect, organize, and produce new data and analyses has provide an impressive resource for marine planners and stakeholders in BC, many of the datasets have exceeded their recommended useful lifespans (Ban et al., 2012). Other datasets such as the plethora composing the commercial fisheries sub-model were collected over 20 years ago with a range of 1991-2010 across the datasets. Moreover, some datasets were collected over a range of years whereas others such as the shipping layers were collected over a single year, and as a result may not be accurate representations of average use.

Different methods utilized, decision making for data processing, along with the inputs to models and rankings represent several limitations. For instance, the use of a MCDA method other than the AHP would likely produce different results. Additionally, the chosen breakdown for the rank-weightings within the weighted linear combination has a significant influence on the results. Human error may have also been present when processing data and creating models. All inherent challenges of academic endeavours though they have been mitigated as much as possible through reanalysis, review, and attention to detail. Another researcher induced limitation is the bias in terms of criteria selected for inclusion and the datasets chosen to represent them.

While many had been used in existing studies and within the MaPP region, others were not included or were proxied. This is of course influenced by the researchers thought processes, existing experiences (work, education, and otherwise), and perhaps most importantly by the literature surveyed.

10.6.2 Stakeholder engagement

As described in section 2.3, stakeholder engagement can be a challenging and imperfect endeavour, yet is one crucial to guiding decision making and ensuring equitable outcomes. The first consideration is related to the scale and timing of the analysis, that is, who should be involved at each stage and scale. For instance, as this study is quite encompassing, those identified for involvement were chosen based on influence over planning processes (e.g. delineating suitable locations for tidal) and those that could represent the multitude of First Nations communities more broadly (e.g. planning conglomerates). While FN planning conglomerates and MaPP senior planners provides proxies for the larger area, it should not be regarded as sufficient engagement for development and does not reflect individual or collective views of the innumerable groups and individuals who might identify as stakeholders. Such is one of the imperfections and challenges of stakeholder engagement, as the temporal and funding constraints of a master's makes it unfeasible to integrate all stakeholders, or even a large subset of them. The Phase II interview was intended to allow for more specific community level engagement, yet it still negated individual contributions from community members. While it may have not been possible within the context of this research to engage at such a level, the utility and need for incorporation of individuals and smaller scale needs, opinions, and values is paramount to the management of common resources and to the development of tidal energy or any RET.

Just as it was deemed unfeasible and unnecessary at the current stage of the tidal industry in BC to engage with all FN communities, there was selective engagement with government. While FLNRO can be perceived as the principal ministry/government agency involved in BC MSP, it represents only one of many ministries at both the provincial and federal level with some sort of regulatory or other oversight regarding tidal energy development. Others include the BC Ministry of Energy, Mines and Petroleum Resources, the DFO, and more in-depth engagement with BC Hydro.

Furthermore, the interdisciplinary nature of the research and its desired function as a starting point for assessing TSTs in the study region necessitated a broad range of questions within the Phase I interview, leading into the Phase II interview touching on a multitude of aspects related to community energy, costs, and future goals. While the collection of encompassing data is paramount to inform decision making, and though the questions were structured to delve into critical questions related to tidal energy development, this range makes the suitability of a single participant to answer all questions unlikely, while HR capacity may limit multiple people from a single organization participating.

Another issue was related to how the interviews were conducted, that is over the phone. The lack of in person engagement risks the loss of data through miscommunication. Moreover, there were issues when trying to engage with stakeholders within the timeframe or those groups that opted not to participate.

Finally, there were significant challenges in establishing communication with a candidate community. This can be partly attributed to the timing of the interview, with contact beginning in late June, as it is a time during which many people take vacations. This, and issues such as HR availability were further compounded by the COVID-19 pandemic.

10.6.3 Scope and scale of the research and interdisciplinary methods

Some of the greatest challenges associated with this research project are derived from interdisciplinary approach used (and necessary) to provide a holistic assessment of tidal energy in the MaPP region. Instead of a single integrated book, this research could have been separated into multiple distinct research projects. However, there are challenges associated with this as well as although these projects could likely be more focused, the quality of data would not have been improved and there may be a lack of integration across the projects. Furthermore, such an endeavour would require additional researchers and the funding to do so. While this research is by no means perfect, the trade off between aspects coverage and level of analysis within each Phase and the interdisciplinary methods used is sufficient as a starting point to guide communities now and recommend future research.

Arising from this integrated framework was the disconnect between identifying the most suitable candidate communities and making a connection to undertake the interview. While ensuring that the socio-economic component and feedback on the identified site was achieved by choosing a candidate community in which an interview could take place, QCV and Skidegate Inlet did not represent the most promising communities for TST development. Thus, the integrated framework ended up being a limitation in that regard as the most economically viable community was evidently not selected as the candidate community

Scale is a critical consideration within this study. Originally the responses of participants were to be averaged for a MaPP regional analysis. However, upon completing the first interview it was clear that each sub-region would have distinct objectives in terms of planning and thus the actual spatial analysis required quadrupled. While this did increase the applicability of the research by making it more location specific, it increased workload and hampered the

researcher's ability to critically analyse results at finer scales.

Chapter XI: Conclusions

This study has developed a framework for assessing the suitability of tidal energy for RCDRC electrification that considers social, economic, and environmental factors/drivers by utilizing integrated methods and knowledge from a range of academic disciplines (e.g. geography, social sciences, economics and engineering). Such a framework heeds the called for paradigm shift from distinct areas of research towards holistic decision making. The results of this study further support the utility of such integrated, multi-disciplinary assessments. The findings summarized in 11.1 along with the future research recommendations in 11.2 provide a solid foundation for potential tidal energy assessments in the MaPP region, BC, and more globally.

11.1   Summary of findings

Energy self-sufficiency was identified as the major driving factor for energy transitions and an important characteristic of future community energy plans. Furthermore, funding and human resource capacity represent two significant and intertwined barriers to both tidal and renewable energy technology development. Tidal energy was favourably viewed within the MaPP region for both community and economic development opportunities, however encompassing information regarding TSTs is of paramount importance in terms of the need for its provision to communities. Environmental impacts are the most important consideration. The results of the GIS-MCDA found substantial tidal resources according to technical feasibility criteria ($\approx$89.8 km^2) in the MaPP region along with identifying suitable areas that mitigate potential conflict with existing uses.

Phase II mapping allowed for the identification of sites and candidate communities (n=9) and a more focused analysis of site suitability by including technical and economic considerations, while also scree. The Phase II interview found that tidal energy would be an acceptable renewable energy technology on Haida Gwaii, with the identified site in Skidegate Inlet (SI) having suitable MSP to be considered a candidate site.

The cost of electricity production for Haida Gwaii's south grid is high at $0.63/kWh (upper bound of $0.91/kWh when considering CO_2 equivalent externalities at a high discount rate and fuel price). This is $0.08-0.14 more per kWh than the literature cited LCOE range for TSTs, suggesting that tidal could be a candidate for baseload renewable energy generation. Despite having a technically viable peak spring current speed, the SI site is financially unviable across all diesel LCOE scenarios, owing to poor capacity factor (1.62%) and high cabling costs due to site-to-community distance.

11.2 Future research recommendations

1. Collaboration between communities (and the public more broadly), government, academia, NGOs and industry to collect, refine, and publish contemporary data to support marine spatial planning efforts in BC; along with updating/maintaining existing datasets related to remote communities.

2. Validation of the Natural Resources Canada tidal resource model at select sites (e.g. Skidegate Inlet site and other sites of interest) and refinement of the model in areas with poor resolution (e.g. Masset Inlet).

3. Refinement of the GIS-MCDA model to identify economically viable sites in the MaPP region based on peak spring current speed greater than 2 m/s and/or

4. selection based on mean peak spring current speed.

5. Develop a roadmap for the creation of a TST test centre on the BC Coast leveraging the resources of groups such as the Pacific Regional Institute for Marine Energy Discovery and Marine Renewables Canada, and foster a community of interested parties (e.g. local communities, industry, academia, government, and non-government organizations).

6. Examination of the economic feasibility, including a more robust cost pathway analysis, of tidal stream energy in other remote coastal diesel reliant First Nations communities. Specifically, Bull Harbor, Gilford Island, and Old Masset.

7. Identification of potential synergistic locations and examination of feasibility for power provision via tidal stream energy for existing/planned remote economic operations on the BC coast.

References

AECOM. (2014). *Tidal Energy: Strategic Environmental Assessment (SEA) Update for the Bay of Fundy.* Halifax: Offshore Energy Environmental Research Association.

Allen, M., Babiker, M., Chen, Y., Connors, S., de Connick, H., van Diemen, R., . . . Zickfeld, K. (2018). *Global Warming of 1.5°C Summary for Policymakers.* Geneva: World Meteorological Organization.

Ali, S., Taweekun, J., Techato, K., Waewsak, J., & Gyawali, S. (2018). GIS based site suitability assessment for wind and solar farms in Songkhla, Thailand. *Renewable Energy, 132*, 1360–1372. https://doi.org/10.1016/j.renene.2018.09.035

Andersson, J., Perez Vico, E., Hammar, L., & Sandén, B. A. (2017). The critical role of informed political direction for advancing technology: The case of Swedish marine energy. *Energy Policy, 101*, 52–64. https://doi.org/10.1016/j.enpol.2016.11.032

Ang, M. R. C. O., Panganiban, I. K., Mamador, C. C., De Luna, O. D. G., Bausas, M. D., & Cruz, J. P. (2016). Development of a multi-device webGIS-based tool for tidal current energy development. *ISPRS Annals of the Photogrammetry, Remote Sensing and Spatial Information Sciences, 3*(8), 65–70. https://doi.org/10.5194/isprs-annals-III-8-65-2016

Arnstein, S. R. (1969). A Ladder of Citizen Participation. *Journal of the American Institute of Planners, 35*(4), 216–224. https://doi.org/10.1080/01944366908977225

Arriaga, M., Brooks, M., & Moore, N. (2016). *Open Access Energy.* Waterloo: Waterloo Global Science Initiative.

Arriaga, M., Canizares, C. A., & Kazerani, M. (2013). Renewable energy alternatives for remote

communities in northern Ontario, Canada. *IEEE Transactions on Sustainable Energy, 4*,

661-670. doi:10.1109/TSTE.2012

Arriaga, M., Canizares, C., & Kazerani, M. (2014). Northern Lights: Access to Electricity in

Canada's Northern and Remote Communities. *Power and Energy Magazine, IEEE 12*, 50-

59.

Balbus, J. M., Greenblatt, J. B., Chari, R., Millstein, D., & Ebi, K. L. (2014). A wedge-based

approach to estimating health co-benefits of climate change mitigation activities in the

United States. *Climatic Change, 127*, 199–210. https://doi.org/10.1007/s10584-014-1262-

5.2234154

Ban, N., & Alder, J. (2008). How wild is the ocean? Assessing the intensity of anthropogenic

marine activities in British Columbia, Canada. *Aquatic Conservation: Marine and

Freshwater Ecosystems, 18*(1), 55–85. https://doi.org/10.1002/aqc.816

Ban, N., Bodtker, K. M., Nicolson, D., Robb, C. K., Royle, K., & Short, C. (2012). Setting the

stage for marine spatial planning: Ecological and social data collation and analyses in

Canada's Pacific waters. *Marine Policy, 39*, 11–20.

https://doi.org/10.1016/j.marpol.2012.10.017

Bank of Canada. (2020). *Inflation Calculator*. Retrieved from Bank of Canada:

https://www.bankofcanada.ca/rates/related/inflation-calculator/

Barrington-Leigh, C., & Ouliaris, M. (2017). The renewable energy landscape in Canada: A spatial analysis. *Renewable and Sustainable Energy Reviews, 75*, 809–819. https://doi.org/10.1016/j.rser.2016.11.061

BC Hydro. (2008, December 9). *BC Hydro Haida Gwaii Draft Request for Proposals Information Session.* Retrieved from BC Hydro: https://www.bchydro.com/content/dam/hydro/medialib/internet/documents/info/pdf/info_-_haida_gwaii7.pdf

BC Hydro. (2012, October 18). *Haida Gwaii - RFEOI.* Retrieved from BC Hydro: https://www.bchydro.com/content/dam/hydro/medialib/internet/documents/planning_regulatory/acquiring_power/2012q4/request_for_expressions.pdf

BC Hydro. (2020, September). Confidential Haida Gwaii Load Data. Vancouver, British Columbia, Canada.

Bedard, R., Jacobson, P. T., Previsic, M., Musial, W., & Varley, R. (2010). An Overview of Ocean Renewable Energy Technologies. *Oceanography, 74*, 22-31.

Black & Veatch. (2020). *Lessons Learnt from MeyGen Phase 1A Final Summary Report.* London: United Kingdom Department for Business, Energy & Industrial Strategy.

Blanchfield, J. B. (2007). *The Extractable Power from Tidal Streams, including a Case Study for Haida Gwaii.* Victoria, Unpublished masters book: University of Victoria.

Bonar, P. A. J., Bryden, I. G., & Borthwick, A. G. L. (2015). Social and ecological impacts of marine energy development. *Renewable and Sustainable Energy Reviews, 47*, 486–495. https://doi.org/10.1016/j.rser.2015.03.068

Boronowski, S. M. (2009). *Integration of Wave and Tidal Power into the Haida Gwaii Electrical Grid*. Victoria, Unpublished Masters book: University of Victoria.

Borthwick, A. G. (2016). Marine Renewable Energy Seascape. *Engineering, 2*, 69-78.

British Columbia Marine Conservation Analysis (BCMCA). (2012). *A Series of Marxan Scenarios for Pacific Canada: a report from the British Columbia Marine Conservation Analysis (BCMCA)*.

BCMCA. (2011). *Marine Atlas of Pacific Canada: A Product of the British Columbia Marine Conservation Analysis (BCMCA)*.

Brosche, P., & Schuh, H. (1998). Tides and Earth Rotation. *Surveys in Geophysics, 19*, 417–430. Retrieved from https://link-springer-com.ezproxy.library.uvic.ca/content/pdf/10.1023%2FA%3A1006515130492.pdf

Bricker, J. D., Esteban, M., Takagi, H., & Roeber, V. (2017). Economic feasibility of tidal stream and wave power in post-Fukushima Japan. *Renewable Energy, 114*, 32–45. https://doi.org/10.1016/j.renene.2016.06.049

Bruck, M., Sandborn, P., & Goudarzi, N. (2018). A Levelized Cost of Energy (LCOE) model for wind farms that include Power Purchase Agreements (PPAs). *Renewable Energy, 122*, 131–139.

Canada's Ocean Supercluster (COS). (2020, July 28). *Ocean Aware Project Announcement*. Retrieved from Canada's Ocean Supercluster: https://oceansupercluster.ca/ocean-aware-project-announcement-news/

Coastal First Nations. (2020, August 3). *Working Toward a 100% Renewable Future*. Retrieved from Coastal First Nations Great Bear Initiative: https://coastalfirstnations.ca/working-toward-a-100-renewable-future/

Coast Funds. (2019, November 5). *Coast Funds and the Province Announce New Fund for First Nations' Investments in Renewable Energy in the Great Bear Rainforest and Haida Gwaii*. Retrieved from Coast Funds: https://coastfunds.ca/news/coast-funds-and-the-province-announce-new-fund-for-first-nations-investments-in-renewable-energy-in-the-great-bear-rainforest-and-haida-gwaii/

Copping, A. (2018). *The State of Knowledge for Environmental Effects*. Lisbon: Ocean Energy Systems.

Copping, A. E., & Hemery, L. G., (Eds.) (2020). *OES-Environmental 2020 State of the Science Report: Environmental Effects of Marine Renewable Energy Development Around the World*. Lisbon: Ocean Energy Systems.

Copping, A. E., Freeman, M. C., Gorton, A. M., & Hemery, L. G. (2020). Risk Retirement-Decreasing Uncertainty and Informing Consenting Processes for Marine Renewable Energy Development. *Marine Science and Engineering, 8*(172), 1–21. https://doi.org/10.3390/jmse8030172

Cradden, L., Kalogeri, C., Martinez Barrios, I., Galanis, G., Ingram, D., & Kallos, G. (2016). Multi-criteria site selection for offshore renewable energy platforms. *Renewable Energy, 87*, 791–806. https://doi.org/10.1016/j.renene.2015.10.035

Cuppen, E., Bosch-Rekveldt, M. G. C., Pikaar, E., & Mehos, D. C. (2016). Stakeholder engagement in large-scale energy infrastructure projects: Revealing perspectives using Q methodology. *International Journal of Project Management, 34,* 1347–1359. https://doi.org/10.1016/j.ijproman.2016.01.003

Dalton, G., Allan, G., Beaumont, N., Georgakaki, A., Hacking, N., Hooper, T., … Stallard, T. (2015). Economic and socio-economic assessment methods for ocean renewable energy: Public and private perspectives, *Renewable and Sustainable Energy Reviews, 45,* 850–878. https://doi.org/10.1016/j.rser.2015.01.068

Dark, S. J., & Bram, D. (2007). The modifi able areal unit problem (MAUP) in physical geography. *Progress in Physical Geography, 31*(5), 471–479. https://doi.org/10.1177/0309133307083294

Davies, I. M., Gubbins, M., & Watret, R. (2012). *Scoping Study for Tidal Stream Energy Development in Scottish Waters.* Edinburgh: The Scottish Government. Retrieved from www.scotland.gov.uk.

Davies, I. M., Watret, R., & Gubbins, M. (2014). Spatial planning for sustainable marine renewable energy developments in Scotland. *Ocean & Coastal Management, 99,* 72-81.

Defne, Z., Haas, K. A., & Fritz, H. M. (2011). GIS based multi-criteria assessment of tidal stream power potential: A case study for Georgia, USA. *Renewable and Sustainable Energy Reviews, 15,* 2310–2321. https://doi.org/10.1016/J.RSER.2011.02.005

De Groot, J., & Bailey, I. (2016). What drives attitudes towards marine renewable energy development in island communities in the UK? *International Journal of Marine Energy, 13*, 80–95. https://doi.org/10.1016/j.ijome.2016.01.007

Deparment of Fisheries and Oceans Canada (DFO). (2018). *Seafisheries Landings*. Retrieved from Fisheries and Oceans Canada: https://www.dfo-mpo.gc.ca/stats/commercial/sea-maritimes-eng.htm

Dijkman, T. J., & Benders, R. M. J. (2010). Comparison of renewable fuels based on their land use using energy densities. *Renewable and Sustainable Energy Reviews, 14*, 3148–3155. https://doi.org/10.1016/j.rser.2010.07.029

Dyer, J., Stringer, L. C., Dougill, A. J., Leventon, J., Nshimbi, M., Chama, F., … Syampungani, S. (2014). Assessing participatory practices in community-based natural resource management: Experiences in community engagement from southern Africa. *Journal of Environmental Management, 137*, 137–145. https://doi.org/10.1016/j.jenvman.2013.11.057

Eidelwein, F., Cisco Collatto, D., Rodrigues, L. H., Pacheco Lacerda, D., & Piran, F. S. (2018). Internalization of environmental externalities: Development of a method for elaborating the statement of economic and environmental results. *Journal of Cleaner Production, 170*, 1316–1327. https://doi.org/10.1016/j.jclepro.2017.09.208

Ehler, C., & Douvere, F. (2009). Marine Spatial Planning: a step-by-step approach toward ecosystem-based management. Paris: UNESCO.

Environment and Climate Change Canada (ECCC). (2016). *National Inventory Report 1990-2014: Greenhouse Gas Sources in Canada.* Gatineau: Government of Canada .

Fitzgerald, E. (2018). *Powering Self-Determination: Indigenous Renewable Energy Developments in British Columbia.* Victoria: Unpublished masters book. University of Victoria.

Frazão Santos, C., Agardy, T., Andrade, F., Crowder, L. B., Ehler, C. N., & Orbach, M. K. (2018). Major challenges in developing marine spatial planning. *Marine Policy.* https://doi.org/10.1016/j.marpol.2018.08.032

Galparsoro, I., Liria, P., Legorburu, I., Bald, J., Chust, G., Ruiz-Minguela, P., Pérez, G., Marqués, J., Torre-Enciso, Y., González, M., & Borja, A. (2012). A Marine Spatial Planning Approach to Select Suitable Areas for Installing Wave Energy Converters (WECs), on the Basque Continental Shelf (Bay of Biscay), *Coastal Management*, 40:1, 1-19, DOI: 10.1080/08920753.2011.637483

Gimpel, A., Stelzenmüller, V., Grote, B., Buck, B. H., Floeter, J., Núñez-Riboni, I., . . . Temming, A. (2015). A GIS modelling framework to evaluate marine spatial planning scenarios: Co-location of offshore wind farms and aquaculture in the German EEZ. *Marine Policy, 55,* 102-115.

Gopnik, M., Fieseler, C., Cantral, L., McClellan, K., Pendleton, L., & Crowder, L. (2012). Coming to the table: Early stakeholder engagement in marine spatial planning. *Marine Policy, 36,* 1139–1149. https://doi.org/10.1016/j.marpol.2012.02.012

Government of British Columbia. (2018). *CleanBC.* Victoria: Government of British Columbia.

Government of British Columbia. (2020). *British Columbia's Carbon Tax*. Retrieved from British

    Columbia: https://www2.gov.bc.ca/gov/content/environment/climate-change/planning-

    and-action/carbon-tax

Government of Canada. (2017, June 21). *Pricing carbon pollution in Canada: how it will work*.

    Retrieved from Government of Canada: https://www.canada.ca/en/environment-climate-

    change/news/2017/05/pricing_carbon_pollutionincanadahowitwillwork.html

Government of Canada. (2019, March 6). *Reducing reliance on diesel*. Retrieved from Reducing

    reliance on diesel:

    https://www.canada.ca/en/services/environment/weather/climatechange/climate-

    plan/reduce-emissions/reducing-reliance-diesel.html

Green Kappa Consulting and Technology Inc. (2018). *Tidal Project Cost Estimation

    Methodology Implemented in Excel*. Ottawa: CanmetENERGY.

Haslett, J. R., Garcia-Llorente, M., Harrison, P. A., Li, S., Berry, P. M., Bugter, R., … Tinch, R.

    (2018). Offshore renewable energy and nature conservation: the case of marine tidal

    turbines in Northern Ireland. *Biodiversity and Conservation*, *27*, 1619–1638.

    https://doi.org/10.1007/s10531-016-1268-6

Hastie, G. D., Russell, D. J. F., Lepper, P., Elliott, J., Wilson, B., Benjamins, S., & Thompson, D.

    (2018). Harbour seals avoid tidal turbine noise: Implications for collision risk. *Journal of

    Applied Ecology*, *55*(2), 684–693. https://doi.org/10.1111/1365-2664.12981

Hatch Energy Ltd. (2008). *Haida Gwaii/Queen Charlotte Islands Demonstration Tidal Power Plant Feasibility Study.* Victoria: BC Ministry of Energy, Mines and Petroleum Resources.

Heiltsuk Tribal Council. (2017). Investigation Report: The 48 hours after the grounding of the Nathan E. Stewart and its oil spill.

Hemer, M. A., Manasseh, R., McInnes, K. L., Penesis, I., & Pitman, T. (2018). Perspectives on a way forward for ocean renewable energy in Australia. *Renewable Energy, 127,* 733–745. https://doi.org/10.1016/j.renene.2018.05.036

Interagency Working Group on Social Cost of Carbon. (2010). *Social Cost of Carbon for Regulatory Impact Analysis under Executive Order 12866.* Washington: Environmental Protection Agency Interagency Working Group on Social Cost of Carbon. https://www.epa.gov/sites/production /files/2016-12/documents/scc_tsd_2010.pdf.

International Renewable Energy Agency (IRENA). (2019). *Renewable Power Generation Costs in 2018.* Abu Dhabi: International Renewable Energy Agency.

IRENA (2020). *Innovation outlook: Ocean energy technologies.* Abu Dhabi: International Renewable Energy Agency.

Irwin, A. (2006). The Politics of Talk: Coming to Terms with the "New" Scientific Governance. Source: *Social Studies of Science,* 36(2), 299–320. https://doi.org/10.1177/0306312706053350

Jahanshahi, A., Kamali, M., Khalaj, M., & Khodaparast, Z. (2019). Delphi-based prioritization of economic criteria for development of wave and tidal energy technologies. *Energy, 167*, 819–827. https://doi.org/10.1016/j.energy.2018.11.040

Janssen, R., Arciniegas, G., & Alexander, K. A. (2014). Decision support tools for collaborative marine spatial planning: identifying potential sites for tidal energy devices around the Mull of Kintyre, Scotland, *58*, 719–737. https://doi.org/10.1080/09640568.2014.887561

Jenkins, L. D., Dreyer, S. J., Polis, H. J., Beaver, E., Kowalski, A. A., Linder, H. L., … Wiesebron, L. E. (2018). Human dimensions of tidal energy: A review of theories and frameworks. *Renewable and Sustainable Energy Reviews, 97*, 323–337. https://doi.org/10.1016/j.rser.2018.08.036

Jenniches, S. (2018). Assessing the regional economic impacts of renewable energy sources – A literature review. *Renewable and Sustainable Energy Reviews, 93*, 35–51. https://doi.org/10.1016/j.rser.2018.05.008

Johannessen, D. I., Macdonald, J. S., Harris, K. A., & Ross, P. S. (2007). *Marine Environmental Quality in the Pacific North Coast Integrated Management Area (PNCIMA), British Columbia, Canada: A Summary of Contaminant Sources, Types, and Risks.* Sidney: Fisheries and Oceans Canada.

Johnson, K., Dalton, G., & Masters, I. (2018). *Building Industries at Sea: 'Blue Growth' and the New Maritime Economy.* Delft: River Publishers.

Karanasios, K., & Parker, P. (2018). Tracking the transition to renewable electricity in remote indigenous communities in Canada. *Energy Policy, 118*, 169–181. https://doi.org/10.1016/j.enpol.2018.03.032

Kennedy, M. (2018). *Energy Shift: Diesel Reliance in Remote Communities in BC.* Vancouver: Unpublished masters book.

Kerr, S., Johnson, K., & Side, J. C. (2014a). Planning at the edge: Integrating across the land sea divide. *Marine Policy, 47,* 118-125.

Kerr, S., Watts, L., Colton, J., Hull, A., Johnson, K., Jude, S., … Vergunst, J. (2014b). Establishing an agenda for social studies research in marine renewable energy. *Energy Policy, 67,* 694–702. https://doi.org/10.1016/j.enpol.2013.11.063

Kerr, S., Colton, J., Johnson, K., & Wright, G. (2015). Rights and ownership in sea country: implications of marine renewable energy for indigenous and local communities. *Marine Policy, 52.* https://doi.org/10.1016/j.marpol.2014.11.002

Kilcher, L., Thresher, R., & Tinnesand, H. (2016). *Marine Hydrokinetic Energy Site Identification and Ranking Methodology Part II: Tidal Energy .* Golden: National Renewable Energy Laboratory.

Kim, C.-K., Toft, J. E., Papenfus, M., Verutes, G., Guerry, A. D., Ruckelshaus, M. H., . . . Polasky, S. (2012). Catching the Right Wave: Evaluating Wave Energy Resources and Potential Compatibility with Existing Marine and Coastal Uses. *PLOS ONE, 7,* 1-14.

Kim, C. K., Jang, S., & Kim, T. Y. (2018). Site selection for offshore wind farms in the southwest coast of South Korea. *Renewable Energy, 120*, 151–162. https://doi.org/10.1016/j.renene.2017.12.081

Knowles, J. (2016). *Power Shift: Electricity for Canada's Remote Communities.* Ottawa: Conference Board of Canada.

Lafleur, C., Truelove, W. A., Cousineau, J., Hiles , C. E., Buckham, B., & Crawford , C. (2020). A screening method to quantify the economic viability of off-grid instream tidal energy deployment. *Renewable Energy, 159*, 610-622.

Lamy, J. V, & Azevedo, I. L. (2018). Do tidal stream energy projects offer more value than offshore wind farms? A case study in the United Kingdom. *Energy Policy, 113*, 28-40. https://doi.org/10.1016/j.enpol.2017.10.030

Lehmann, P., Sijm, J., Gawel, · Erik, Strunz, S., Chewpreecha, U., Mercure, J.-F., & Pollitt, H. (2018). Addressing multiple externalities from electricity generation: a case for EU renewable energy policy beyond 2020? *Environmental Economics and Policy Studies.* https://doi.org/10.1007/s10018-018-0229-6

Lewis, M., Neill, S. P., Robins, P. E., & Hashemi, M. R. (2015). Resource assessment for future generations of tidal-stream energy arrays. *Energy, 83*, 403–415.

LiVecchi, A., A. Copping, D. Jenne, A. Gorton, R. Preus, G. Gill, R. Robichaud, R. Green, S. Geerlofs, S. Gore, D. Hume, W. McShane, C. Schmaus, H. Spence. 2019. *Powering the Blue Economy; Exploring Opportunities for Marine Renewable Energy in Maritime*

Markets. U.S. Department of Energy, Office of Energy Efficiency and Renewable Energy. Washington, D.C.

Lohmann, L. (2009). Toward a different debate in environmental accounting: The cases of carbon and cost–benefit. *Accounting, Organizations and Society, 34*, 499–534.

MacDougall, S., Colton, J., Daborn, G., & Karsten, R. (2013). *Community and business toolkit for tidal energy development*. Wolfville: Acadia Tidal Energy Institute.

MacDougall, S. L. (2017). Confronting the financing impasse: Risk management through internationally staged investments in tidal energy development. *International Journal of Marine Energy, 18*, 78–87. https://doi.org/10.1016/j.ijome.2017.03.002

Magagna, D. (2019). *Low Carbon Energy Observatory: Ocean Energy Technology Development Report*. Luxembourg: European Commission.

Marine Planning Partnership (MaPP) Initiative. (2015a). *Central Coast Marine Plan*.

Marine Planning Partnership (MaPP) Initiative. (2015b). *Haida Gwaii Marine Plan*.

Marine Planning Partnership (MaPP) Initiative. (2015c). *North Coast Marine Plan*.

Marine Planning Partnership (MaPP) Initiative. (2015d). *North Vancouver Island Marine Plan*.

Marine Planning Partnership (MaPP) Initiative. (2016). *Regional Action Framework*.

Marine Renewables Canada. (2018). *Marine Renewable Energy in Canada: State of the Sector Report*. Halifax: Marine Renewables Canada.

Maslov, N., Brosset, D., Claramunt, C., Charpentier, J.-F., Maslov, N., Brosset, D., … Charpentier, J.-F. (2014). A Geographical-Based Multi-Criteria Approach for Marine Energy Farm Planning. *ISPRS International Journal of Geo-Information*, *3*(2), 781–799. https://doi.org/10.3390/ijgi3020781

Mavi Innovations. (2017). *A pre-feasibility study for river hydrokinetic power projects on the Fraser River-Confidential Report*. Mavi Innovations.

McFarlan, A. (2018). Techno-economic assessment of pathways for electricity generation in northern remote communities in Canada using methanol and dimethyl ether to replace diesel. *Renewable and Sustainable Energy Reviews*, *90*, 863–876. https://doi.org/10.1016/j.rser.2018.03.076

Ministry of Energy, Mines and Petroleum Resources (MEMPR). (2020). *Comprehensive Review of BC Hydro: Phase 2 Interim Report*. Vancouver: Government of British Columbia.

Ministry of Environment and Climate Change Strategy (MECCS). (2019). *2018 B.C. Methodological Guidance for Quantifying Greenhouse Gas Emissions*. Victoria: Ministry of Environment and Climate Change Strategy.

Murali, K., & Sundar, V. (2015). Reassessment of tidal energy potential in India and a decision-making tool for tidal energy technology selection. *International Journal of Ocean and Climate Systems*, *8*(2), 85–97. https://doi.org/10.1177/1759313117694629

Nash, S., & Phoenix, A. (2017). A review of the current understanding of the hydro-environmental impacts of energy removal by tidal turbines. *Renewable and Sustainable Energy Reviews, 80*, 648–662. https://doi.org/10.1016/j.rser.2017.05.289

Natural Resources Canada (NRCan). (2011). *Status of Remote/Off-Grid Communities in Canada.* Government of Canada.

NRCan. (2013). *The First Canadian Smart Remote Microgrid: Hartley Bay, BC.* The Government of Canada.

NRCan. (2019). *Energy Future Announcement Backgrounder*. Retrieved from Government of Canada: https://www.canada.ca/en/natural-resources-canada/news/2019/05/energy-future-announcement-backgrounder.html

NRCan. (2020a). *Monthly Average Retail Prices for Diesel*. Retrieved from Natural Resources Canada: http://www2.nrcan.gc.ca/eneene/sources/pripri/prices_bycity_e.cfm?productID=5&locationID=66&locationID=3&frequency=M&priceYear=2020&Redisplay=

NRCan. (2020b). *Photovoltaic potential and solar resource maps of Canada*. Retrieved from Government of Canada: https://www.nrcan.gc.ca/energy-efficiency/data-research-insights-energy-ef/buildings-innovation/solar-photovoltaic-energy-buildi/resources/photovoltaic-potential-and-solar-resource-maps-canada/18366

Nissen, U., & Harfst, N. (2019). Shortcomings of The Traditional "Levelized Cost of Energy" [LCOE] for The Determination of Grid Parity. *Energy, 171.* https://doi.org/10.1016/j.energy.2019.01.093

Nova Scotia Department of Energy and Mines (N.S. DEM). (2011). *Marine Renewable Energy Infrastructure Assessment.* Halifax: Nova Scotia Department of Energy and Mines.

Ocean Energy Systems (OES). 2015. *International Levelized Cost of Energy for Ocean Energy Technologies.* Lisbon: International Energy Agency.

OES. (2019). *An Overview of Ocean Energy Activities in 2019.* Lisbon: International Energy Agency.

OES. (2020), *Ocean Energy in Islands and Remote Coastal Areas: Opportunities and Challenges.* Lisbon: International Energy Agency.

O'Carroll, J. P. J., Kennedy, R. M., Creech, A., & Savidge, G. (2017). Tidal Energy: The benthic effects of an operational tidal stream turbine. *Marine Environmental Research, 129,* 277–290. https://doi.org/10.1016/j.marenvres.2017.06.007

O'Rourke, F. O., Boyle, F., & Reynolds, A. (2009). Tidal energy update 2009. *Applied Energy, 87,* 398–409. https://doi.org/10.1016/j.apenergy.2009.08.014

O'Rourke, F., Boyle, F., & Reynolds, A. (2010a). Marine current energy devices: Current status and possible future applications in Ireland. *Renewable and Sustainable Energy Reviews, 14,* 1026–1036. https://doi.org/10.1016/j.rser.2009.11.012

O'Rourke, F., Boyle, F., & Reynolds, A. (2010b). Tidal current energy resource assessment in Ireland: Current status and future update. *Renewable and Sustainable Energy Reviews*, *14*, 3206–3212. https://doi.org/10.1016/j.rser.2010.07.039

Oxera. (2011). *Discount rates for low-carbon and renewable generation technologies.* Oxford: Oxera Consulting Ltd.

Pacific North Coast Integrated Management Area (PNCIMA) Initiative. (2017). *Pacific North Coast Integrated Management Area Plan.*

Pine, M. K., Schmitt, Pál., Culloch, R. M., Lieber, L., & Kregting, L. T. (2019). Providing ecological context to anthropogenic subsea noise: Assessing listening space reductions of marine mammals from tidal energy devices. *Renewable and Sustainable Energy Reviews*, *103*, 49–57.

Polis, H. J., Dreyer, S. J., & Jenkins, L. D. (2017). Public willingness to pay and policy preferences for tidal energy research and development: A study of households in Washington State. *Ecological Economics*, *136*, 213–225. https://doi.org/10.1016/j.ecolecon.2017.01.024

Ponsoni, L., Nichols, C., Buatois, A., Kenkhuis, J., Schmidt, C., de Haas, P., … Nauw, J. J. (2018). Deployment of a floating tidal energy plant in the Marsdiep inlet: resource assessment, environmental characterization and power output. *Journal of Marine Science and Technology*, 1–16. https://doi.org/10.1007/s00773-018-0590-y

Pomeroy, R., & Douvere, F. (2008). The engagement of stakeholders in the marine spatial planning process. *Marine Policy*, 32, 816–822. https://doi.org/10.1016/j.marpol.2008.03.017

Price Waterhouse Cooper. (2015). *Wataynikaneyap Power Project: Socioeconomic Impact Analysis of Building Grid Connection to Ontario's Remote Communities.* Wataynikaneyap Power Project.

Province of British Columbia. (2018). *cleanBC: our nature. our power. our future.* Victoria: Province of British Columbia.

Quero García, P., García Sanabria, J., & Chica Ruiz, J. A. (2019). The role of maritime spatial planning on the advance of blue energy in the European Union. *Marine Policy*, *99*, 123–131. https://doi.org/10.1016/j.marpol.2018.10.015

Reed, M. S. (2008). Stakeholder participation for environmental management: A literature review. *Biological Conservation*, 141, 2417–2431. https://doi.org/10.1016/j.biocon.2008.07.014

Rezai, A., Foley, D. K., & Taylor, L. (2012). Global warming and economic externalities. *Economic Theory*, *49*, 329–351. https://doi.org/10.1007/s00199-010-0592-4

Rezaei, M., & Dowlatabadi, H. (2016). Off-grid: community energy and the pursuit of self-sufficiency in British Columbia's remote and First Nations communities. *Local Environment*, *21*, 789–807. https://doi.org/10.1080/13549839.2015.1031730

Richardson, R. L. (2018). *A Comparative Analysis of Marine Spatial Planning and Policies for Marine Renewable Energy in Scotland and British Columbia*. Victoria: Unpublished honours book.

Ricke, K., Drouet, L., Caldeira, K., & Tavoni, M. (2018). Country-level social cost of carbon. *Nature Climate Change*, 8, 895–900. https://doi.org/10.1038/s41558-018-0282-y

Ritchie, H., & Ellis, G. (2010). Exploring Stakeholder Engagement in Marine Spatial Planning. *Journal of Environmental Planning and Management*, 53(6), 701–723. https://doi.org/10.1080/09640568.2010.488100

Roberts, A., Thomas, B., Sewell, P., Khan, Z., Balmain, S., & Gillman, J. (2016). Current tidal power technologies and their suitability for applications in coastal and marine areas. *Journal of Ocean Engineering and Marine Energy*, 2, 227–245. https://doi.org/10.1007/s40722-016-0044-8

Robertson, B., Bailey, H., & Buckham, B. (2017). *Wave Energy: A Primer for British Columbia*. Victoria: Pacific Institute for Climate Solutions.

Rowe, G., & Frewer, L. J. (2000). Public Participation Methods: A Framework for Evaluation. *Science, Technology, & Human Values*, 25(1), 3–29. Retrieved from https://www-jstor-org.ezproxy.library.uvic.ca/stable/pdf/690198.pdf?refreqid=excelsior%3A00b24687ce15 d89717abb79bc8b86e72

Rowe, G., & Frewer, L. J. (2005). A Typology of Public Engagement Mechanisms. *Science, Technology, & Human Values*, 30(2), 251–290. Retrieved from https://journals-sagepub-com.ezproxy.library.uvic.ca/doi/pdf/10.1177/0162243904271724

Roy, A., Auger, F., Dupriez-Robin, F., Bourguet, S., & Tran, Q. T. (2018). Electrical power supply of remote maritime areas: A review of hybrid systems based on marine renewable energies. *Energies, 11*(7). https://doi.org/10.3390/en11071904

Ruano-Chamorro, C., Carlos Castilla, J., & Gelcich, S. (2018). Human dimensions of marine hydrokinetic energies: Current knowledge and research gaps. *Renewable and Sustainable Energy Reviews, 82*, 1979–1989. https://doi.org/10.1016/j.rser.2017.07.023

San Cristóbal, J. R. (2011). Multi-criteria decision-making in the selection of a renewable energy project in Spain: The vikor method. Renewable Energy, 36(2), 498-502. doi:10.1016/j.renene.2010.07.031

Sangiuliano, S. J. (2017a). Planning for tidal current turbine technology: A case study of the Gulf of St. Lawrence. *Renewable and Sustainable Energy Reviews, 70*, 805–813. https://doi.org/10.1016/j.rser.2016.11.261

Sangiuliano, S. J. (2017b). Turning of the tides: Assessing the international implementation of tidal current turbines. *Renewable and Sustainable Energy Reviews, 80*, 971–989. https://doi.org/10.1016/j.rser.2017.05.045

Segura, E., Morales, R., & Somolinos, J. A. (2017a). Cost assessment methodology and economic viability of tidal energy projects. *Energies, 10* (11), 1–27. https://doi.org/10.3390/en10111806

Segura, E., Morales, R., Somolinos, J. A., & López, A. (2017b). Techno-economic challenges of tidal energy conversion systems: Current status and trends. *Renewable and Sustainable Energy Reviews, 77*, 536–550. https://doi.org/10.1016/j.rser.2017.04.054

Segura, E., Morales, R., & Somolinos, J. A. (2018). A strategic analysis of tidal current energy conversion systems in the European Union. *Applied Energy, 212*, 527–551. https://doi.org/10.1016/j.apenergy.2017.12.045

Segura, E., Morales, R., & Somolinos, J. A. (2019). Increasing the Competitiveness of Tidal Systems by Means of the Improvement of Installation and Maintenance Maneuvers in First Generation Tidal Energy Converters—An Economic Argumentation. *Energies, 12*, 1–27. https://doi.org/10.3390/en12132464

Sen, S., & Ganguly, S. (2017). Opportunities, barriers and issues with renewable energy development – A discussion. *Renewable and Sustainable Energy Reviews, 69*, 1170–1181. https://doi.org/10.1016/j.rser.2016.09.137

Sheppard, E. (2015) Thinking geographically: globalizing capitalism and beyond. *Annals of Association of American Geographers, 105*, 1113-1134

Shucksmith, R. J., & Kelly, C. (2014). Data collection and mapping – Principles, processes and application in marine spatial planning, *Marine Policy 50*, 27–33. https://doi.org/10.1016/j.marpol.2014.05.006

Smart, G., & Noonan, M. (2018). *Tidal Stream and Wave Energy Cost Reduction and Industrial Benefit*. Glasgow: Offshore Renewable Energy Catapult.

Sovacool, B. K., Martiskainen, M., Hook, A., & Baker, L. (2020). Beyond cost and carbon: The multidimensional co-benefits of low carbon transitions in Europe. *Ecological Economics, 169*, 106529. https://doi.org/10.1016/j.ecolecon.2019.106529

Statistics Canada. (2016a). *Queen Charlotte Village 2016 Census Profile*. Retrieved from Statistics Canada: https://www12.statcan.gc.ca/census-recensement/2016/dp-pd/prof/details/Page.cfm?Lang=E&Geo1=CSD&Code1=5947026&Geo2=PR&Code2=59&SearchText=Skidegate&SearchType=Begins&SearchPR=01&B1=All&type=0

Statistics Canada. (2016b). *Skidegate 2016 Census Profile*. Retrieved from Statistics Canada: https://www12.statcan.gc.ca/census-recensement/2016/dp-pd/prof/details/page.cfm?Lang=E&Geo1=CSD&Code1=5947804&Geo2=CD&Code2=5947&SearchText=Skidegate&SearchType=Begins&SearchPR=01&B1=All&TABID=1&type=0

Stern, N. (2007). *The economics of climate change: The Stern review*. Cambridge: Cambridge University Press.

Stillwell, W.G., Seaver, D.A., Edwards, W. (1981). A Comparison of Weight Approximation Techniques in Multiattribute Utility Decision-Making. *Organizational Behavior and Human Performance, 28*, 62-77.

Stothers, R., Klaptocz, V. (2016). *Market Opportunity for Developing Projects in Canada using Marine Hydrokinetic Turbines-Confidential Report*. Mavi Innovations.

Strategic Energy Technologies Information System (SETIS). (2019). *SETIS Magazine: Ocean Energy*. Luxembourg: Publications Office of the European Union.

Streimikiene, D., Balezentis, T., Alisauskaite-Seskiene, I., Stankuniene, G., & Simanaviciene, Z. (2019, April 18). A review of willingness to pay studies for climate change mitigation in the energy sector. *Energies*. MDPI AG. https://doi.org/10.3390/en12081481

Sustainable Marine Energy Ltd. (n.d.). *PLAT-I tidal energy platform*. Retrieved from Sustainable Marine Energy: https://s3-eu-west-1.amazonaws.com/assets-sustainablemarine-com/downloads/SME_PLAT-I-Web-Brochure.pdf

Swiilawiid. (n.d.). *Energy on Haida Gwaii*. Retrieved from Swiilawiid Sustainability Society: http://www.swiilawiid.org/energy-on-haida-gwaii

Tasker, J. P. (2020, December 11). *Ottawa to hike federal carbon tax to $170 a tonne by 2030* . Retrieved from CBC: https://www.cbc.ca/news/politics/carbon-tax-hike-new-climate-plan-1.5837709

Tawil, T. El, Frédéric Charpentier, J., & Benbouzid, M. (2018). Sizing and rough optimization of a hybrid renewable-based farm in a stand-alone marine context. *Renewable Energy, 115,* 1134-1143. https://doi.org/10.1016/jrenene.2017.08.093

Triton Consultants Ltd. (2002). *Tidal Current Energy.* Vancouver: Triton Consultants Ltd.

Truelove, A. (2020). *Updated Viability Logic for Tidal Stream*. Victoria: Integrated Energy Systems Victoria.

Uihlein, A., & Magagna, D. (2016). Wave and tidal current energy - A review of the current state of research beyond technology. *Renewable and Sustainable Energy Reviews*, *58*, 1070–1081. https://doi.org/10.1016/j.rser.2015.12.284

United Nations Environment Programme (UNEP) (2019). Emissions Gap Report 2019. Executive summary. United Nations Environment Programme, Nairobi.

Valleau, N. (2020, March 17). *Impact to Alberta's oil and gas could put Canada on brink of recession, says Conference Board*. Retrieved from Canadian Broadcasting Corporation: https://www.cbc.ca/news/canada/calgary/alberta-oil-gas-covid-19-opec-russia-economy-recession-1.5500733

Van Cleve, F.B., Judd, C., Radil, A., Ahmann, J., Geerlofs, S.H., (2013). *Geospatial Analysis of Technical and Economic Suitability for Renewable Ocean Energy Development on Washington's Outer Coast* . Richland: Pacific Northwest National Laboratory .

van der Ploeg, F., & de Zeeuw, A. (2019). Pricing Carbon and Adjusting Capital to Fend Off

Climate Catastrophes. *Environmental and Resource Economics, 72*, 29–50.

https://doi.org/10.1007/s10640-018-0231-2

van Hees, S. (2019). Increased integration between innovative ocean energy and the EU habitats,

species and water protection rules through Maritime Spatial Planning. *Marine Policy,

100*, 27–42.

Vasileiou, M., Loukogeorgaki, E., & Vagiona, D. G. (2017). GIS-based multi-criteria decision

analysis for site selection of hybrid offshore wind and wave energy systems in Greece.

Renewable and Sustainable Energy Reviews, 73, 745–757.

https://doi.org/10.1016/j.rser.2017.01.161

Vazquez, A., & Iglesias, G. (2015). LCOE (levelized cost of energy) mapping: A new geospatial

tool for tidal stream energy. *Energy, 91*, 192–201.

https://doi.org/10.1016/J.ENERGY.2015.08.012

Vazquez, A., & Iglesias, G. (2016a). A holistic method for selecting tidal stream energy hotspots

under technical, economic and functional constraints. *Energy Conversion and

Management, 117*, 420–430. https://doi.org/10.1016/j.enconman.2016.03.012

Vazquez, A., & Iglesias, G. (2016b). Capital costs in tidal stream energy projects: A spatial

approach. *Energy, 107*. https://doi.org/10.1016/j.energy.2016.03.123

Vazquez, A., & Iglesias, G. (2016c). Grid parity in tidal stream energy projects: An assessment

of financial, technological and economic LCOE input parameters. *Technological*

Forecasting and Social Change, *104*, 89–101.
https://doi.org/10.1016/J.TECHFORE.2015.12.007

Villate, J. L., Pirttimaa, L., & Cochrane, C. (2020). *Strategic Research and Innovation Agenda for Ocean Energy*. Antwerp: European Technology & Inovation Platform for Ocean Energy.

Wang, J.-J., Jing, Y.-Y., Zhang, C.-F., & Zhao, J.-H. (2009). Review on multi-criteria decision analysis aid in sustainable energy decision-making. *Renewable and Sustainable Energy Reviews*, *13*, 2263–2278. https://doi.org/10.1016/j.rser.2009.06.021

Webler, T., & Tuler, S. (2002). Unlocking the Puzzle of Public Participation. *Bulletin of Science, Technology & Society*, 22(3), 179–189. Retrieved from https://journals-sagepub-com.ezproxy.library.uvic.ca/doi/pdf/10.1177/02767602022003002

Whitman, G. P., Pain, R., & Milledge, D. G. (2015). Going with the flow? Using participatory action research in physical geography. *Progress in Physical Geography*, 39(5), 622–639. https://doi.org/10.1177/0309133315589707

Wright, G. (2015). Marine governance in an industrialised ocean: A case study of the emerging marine renewable energy industry. *Marine Policy*, *52*, 77-84.

Xu, X., (2018). *Identifying British Columbia's Strategically Important Wave Energy Sites*. Victoria: Unpublished master's book.

Zaunbrecher, B. S., Daniels, B., Roß-Nickoll, M., & Ziefle, M. (2018). The social and ecological footprint of renewable power generation plants. Balancing social requirements and ecological impacts in an integrated approach. *Energy Research & Social Science, 45,* 91–106. https://doi.org/10.1016/j.erss.2018.07.015

Appendix A: Human Research Ethics Certification

Office of Research Services | Human Research Ethics Board
Michael Williams Building Rm B202 PO Box 1700 STN CSC Victoria BC V8W 2Y2 Canada
T 250-472-4545 | F 250-721-8960 | uvic.ca/research | ethics@uvic.ca

Certificate of Approval

PRINCIPAL INVESTIGATOR	Rosaline Canessa (Supervisor)	ETHICS PROTOCOL NUMBER	19-0327
		Expedited review - delegated	
PRINCIPAL APPLICANT	Riley Richardson Master's student	ORIGINAL APPROVAL DATE	27-Sep-2019
UVIC DEPARTMENT	Geography	APPROVED ON	27-Sep-2019
		APPROVAL EXPIRY DATE	26-Sep-2020

PROJECT TITLE Investigating the Spatial, Social, and Economic Feasibility of Tidal Energy in British Columbia's Remote Coastal Diesel Reliant Communities

RESEARCH TEAM MEMBERS
Brad Buckham - Committee member, UVic
Lauren McWhinnie - Committee member, UVic

DECLARED PROJECT FUNDING
Natural Sciences and Engineering Research Council, Institute for Integrated Energy Systems
Western Economic Diversification Canada, Institute for Integrated Energy Systems

DOCUMENTS INCLUDED IN THIS APPROVAL
Group 2 Recruitment Script.docx - 23-Sep-2019
Group 1 Recruitment Script.docx - 23-Sep-2019
Participant Consent Form.doc - 23-Sep-2019
Draft interview questions.docx - 26-Jul-2019

CONDITIONS OF APPROVAL

This Certificate of Approval is valid for the above term provided there is no change in the protocol.

Modifications
To make any changes to the approved research procedures in your study, please submit a "Request for Modification" form. You must receive ethics approval before proceeding with your modified protocol.

Renewals
Your ethics approval must be current for the period during which you are recruiting participants or collecting data. To renew your protocol, please submit a "Request for Renewal" form before the expiry date on your certificate. You will be sent an emailed reminder prompting you to renew your protocol about six weeks before your expiry date.

Project Closures
When you have completed all data collection activities and will have no further contact with participants, please notify the Human Research Ethics Board by submitting a "Notice of Project Completion" form.

Certification

This certifies that the UVic Human Research Ethics Board has examined this research protocol and concluded that, in all respects, the proposed research meets the appropiate standards of ethics as outlines by the University of Victoria Research Regulations Involving Human Participants.

Dr. Rachael Scarth
Associate VP Research Operations

Certificate issued On: 27-Sep-2019

University of Victoria

Participant Consent Form

Investigating the Spatial, Social, and Economic Feasibility of Tidal Energy in British Columbia's Remote Coastal Diesel Reliant Communities

You are invited to participate in a study entitled "Investigating the Feasibility of Tidal Energy in British Columbia's Remote Coastal Diesel Reliant Communities" that is being conducted by Riley Richardson.

Riley Richardson is a master's student in the department of Geography at the University of Victoria, and you may contact him if you have further questions by email

As a Graduate student, I am required to conduct research as part of the requirements for a Master of Science in Geography degree. It is being conducted under the supervision of Dr. Rosaline Canessa. You may contact my supervisor by email

This research is being funded by the Natural Sciences and Engineering Research Council of Canada, The University of Victoria, and Western Economic Diversification Canada

Purpose and Objectives
The purpose of this research project is to determine suitable locations for tidal energy development in British Columbia based on resource potential along with environment, social, and economic considerations. Furthermore, it seeks to investigate the economic feasibility of tidal energy within an identified remote diesel reliant community.

Importance of this Research
As our window to keep global temperature increase below 1.5°C and mitigate the effects of climate change rapidly closes, the need for a global transition to renewables is increasingly critical. This research will develop a framework for tidal energy suitability analysis, while also helping remote coastal communities to transition off diesel and onto geographically abundant resources such as tidal energy. This research will also add to existing Marine Spatial Planning efforts in BC while protecting the environment, social values, and existing/planned economic opportunities. Such research will empower remote communities with the necessary information and understanding of renewable alternatives, such as tidal, and aid in their planning/decision efforts in determining the most sustainable and viable pathway forward for their communities.

Participants Selection
You are being asked to participate in this study because you are a representative/employee of one of the following groups and were involved in/have knowledge of the Marine Plan Partnership for the Pacific North Coast (MaPP) planning efforts: the Central Coast Indigenous Resource Alliance, the Council of the Haida Nation, the Nanwakolas Council, the North Coast-Skeena First Nations Stewardship Society, or the Ministry of Forests, Lands, Natural Resource Operations and Rural Development.

What is involved
If you consent to voluntarily participate in this research, your participation will include an approximately 45-minute interview answering a set of questions with the primary researcher, Riley Richardson. The interview will take place either over the phone, in person, or over Skype in early 2020.

Subject to your approval you will be audio recorded to facilitate note taking. These audio files are strictly for maintaining accuracy and will only be used by the primary researcher. In addition, handwritten notes will be taken. If you do not wish to be recorded, please inform the primary researcher prior to the interview.

Inconvenience
The only potential inconvenience of participating in this study is the time required to do so.

Risks
There are no known or anticipated risks to you by participating in this research.

Benefits
Participants will have the opportunity collaborate and help shape the development of tidal energy suitability mapping within the study area. The identified community will benefit from a holistic and interdisciplinary feasibility analysis of tidal energy integration within their community encompassing social, economic, and environment considerations.

Society will benefit on local, provincial, national, and international levels. At the local level, this study represents the opportunity for assessing development of an emerging renewable energy source to mitigate the negative externalities associated with diesel generated electricity such as health impacts, high costs of electricity, and constraints on development.

Furthermore, it will provide the potential for new economic opportunities along with providing the community with enhanced understanding of potential pathways forward, enhancing their ability to make decisions for the future. Provincially, this research will help achieve emission reduction targets along with highlighting the feasibility of an emerging industry. Federally, this project will aid in achieving climate change mitigation targets, build upon Natural Resources Canada electrification and technology development objectives, along with building upon Canada's contribution to the state of Marine Renewable Energy knowledge. Internationally this project will help mitigate climate change, provide an integrated framework for assessing tidal suitability for use in other regions, and contribute to the existing tidal energy socio-economic and spatial suitability academic literature.

Voluntary Participation
Your participation in this research must be completely voluntary. If you do decide to participate, you may withdraw at any time without explanation or any consequence. If you do withdraw from the study your data will still be used, subject to your consent. If you do not wish for your data to be used it will be destroyed. To withdraw from the study, please contact Riley Richardson.

Anonymity
In terms of protecting your anonymity, you will be kept anonymous in the distribution of the results. However, contingent upon your consent your organization and/or role (e.g. Planner) will be included, and thus it is possible that members of the public would be able to identify you.

Confidentiality
Your confidentiality and the confidentiality of the data will be protected by storing the hard copy interview notes within a combination locked safe, and the electronic data within password protected computer files. The primary researcher and his supervisors will have access to the data.

There are limits to confidentiality due to the small number of First Nation resource planning alliances within British Columbia. Steps are being taken to prevent this, by removing identifying information of the participants other than their organization (or if consent is not given for dissemination of their organization simply by stating their general role e.g. "government official").

Dissemination of Results
It is anticipated that the results of this study will be shared with others in the following ways: Master of Science thesis and thesis dissertation, online copy of the thesis, presentations at scholarly meetings and conferences, and within at least one (or more) published paper(s).

Disposal of Data
Data from this study will be likely used by the Pacific Regional Institute for Marine Energy Discovery and First Nation communities for planning and research related to the integration of tidal energy in remote communities. It is not known at this time the duration of data use. Data will be stored within a password protected computer(s). When appropriate, electronic data will be destroyed by deleting the files (interview responses and audio tapes) and paper data (field notes, copies of interviews) will be shredded and properly disposed of.

Contacts
Individuals that may be contacted regarding this study include:

Riley L Richardson	Dr. Rosaline Canessa	Dr. Brad Buckham
Primary Researcher	Supervisor	Committee member

In addition, you may verify the ethical approval of this study, or raise any concerns you might have, by contacting the Human Research Ethics Office at the University of Victoria (250-472-4545 or ethics@uvic.ca).

Your signature below indicates that you understand the above conditions of participation in this study, that you have had the opportunity to have your questions answered by the researchers, and that you consent to participate in this research project.

_______________________________   _______________________________   _______________

Name of Participant   *Signature*   *Date*

I consent to be identified by name / credited in the results of the study: _______________ (Participant to provide initials)

A copy of this consent will be left with you, and a copy will be taken by the researcher.

Appendix B: Analytic Hierarchy Process Constraints, Criteria, and Sensitivity Analysis Methods

B.1   AHP criteria and constraints

Table B.1: Tidal stream turbine site constraints.

Constraints on tidal stream turbine site feasibility	Rationale for inclusion	Optimisation	Source layer (see Appendix C)
Peak Spring current speed	Peak spring tidal current speed and mean peak spring tidal current are two frequently cited metrics for assessing the feasibility of a tidal site. Within this study, peak spring tidal current speed was taken as a resource feasibility and suitability metric based on data availability. The literature suggests lower bounds on peak spring current speed of 1.2 m/s (AECOM, 2014) and 1.5 m/s (O'Rourke et al., 2010b; Segura et al., 2017b; Vazquez & Iglesias, 2016a). High potential sites (e.g. those that would be technically and potentially economically viable) can be found at locations where the peak spring tidal current speed exceeds 2 m/s (Nash & Phoenix, 2017; O'Rourke et al., 2010a), 2.25 m/s (Roberts et al., 2016) and 2.5 m/s (Lewis et al., 2015; Segura et al., 2017b).	For site technical feasibility, all locations with spring peak tidal current speed greater or equal to 1.5 m/s will be considered. Locations with peak spring tidal current speed less than 1.5 m/s are not feasible and are constrained.	1
Depth	Depths that are too shallow will hamper device operation, while increasing depth in turn increases the cost and technical challenges of deploying TST devices. Aside from sufficient tidal resource, depth is the main physical parameter in determining a sites suitability (Cradden et al., 2016). The literature ranges in terms of depth with a values of >5 m, 10-40 m, 10-80 m, 20-40 m, less than 40 m, 25-50 m, and 40 to 60 m (AECOM, 2014; Black & Veatch, 2005; Defne et al., 2011; Gahan, 2012; Murali & Sundar, 2015; Roberts et al., 2016; Segura et al., 2019). Others suggest 250 m as being the maximum operable depth for floating MRE platforms, with a	As this study is assessing feasibility based on a generic tidal device which may be bottom mounted or floating, a depth range of 5-100 m will be classified as suitable, with depths less than 5m and greater than 100 m being bounding constraints. Locations where depths are between 5-100 m are suitable, locations where depths are less than 5 m or greater than 100 m are	33

	caveat that 100 m might be considered the current operable limit (Cradden et al., 2016).	unsuitable and constrained.	
Tenured areas	Tenured areas represent a legal ownership of the marine space, which has the potential to exclude other marine uses such as tidal energy.	Areas with an existing tenure are taken as constrained areas.	31
Protected areas and parks	These areas have high conservation value and are protected under federal, provincial, or municipal laws.	Protected areas and parks are constrained areas.	32

Criteria used within a MCDA should reflect the required needs of the energy system (e.g. sufficient resource), be consistent with decision making objectives (e.g. existing MSP), be independent of other criteria, be capable of being quantitatively or qualitatively expressed, and be comparable to other criteria (usually achieved through normalization) (Wang et al., 2009). Table B.2 below details the selected criteria.

Table B.2: Overview of tidal stream energy site criteria.

Criteria to assess tidal stream turbine site suitability	Rationale for inclusion	Optimisation	Source layer (see table 5)
Peak Spring current speed	The suitability of a potential site is dependent on sufficient resource potential, of which peak spring current speed is a good indicator of technical feasibility.	Peak spring tidal current speed ≥ 1.5 m/s as lower bound, see 4.3.1 for distribution.	1
Marine conservation value	The study region is home to a diverse and rich array of marine life essential to maintaining ecological productivity. TST site suitability is dependent upon having minimal impact on existing environmental and ecological uses.	Areas with high conservation value are less suitable, areas with low conservation value are more suitable.	2
Commercial fisheries catch value	Commercial fisheries represent a substantial portion of the economy within the study region. Catch value provides an approximation of which areas have historically been most valuable to the commercial fisheries sector.	Areas with higher catch value are less suitable, areas with lower catch value are more suitable	3-16, 18, 19
Government and research vessel traffic hours	Government and research activities occur within the study region. Areas with high traffic may increase conflict between these vessels and potential tidal development.	Areas with high traffic hours are less suitable than areas with low traffic hours	17

Merchant vessel traffic hours	With three large ports in the study region there is merchant vessel traffic in the area that may potentially conflict with tidal energy.	Areas with high traffic hours are less suitable than areas with low traffic hours	17
Passenger and cruise vessel traffic hours	There are several ferry and seasonal cruise vessel routes within the study region that may overlap with tidal energy sites.	Areas with high traffic hours are less suitable than areas with low traffic hours	17
Tanker vessel traffic hours	Oil and natural gas exportation occur in the study region with proposed expansions which might potentially conflict with tidal energy.	Areas with high traffic hours are less suitable than areas with low traffic hours	17
Tug and service vessel traffic hours	Tug and service vessel operations are critical for several resource extractive activities such as log-towing and goods transportation.	Areas with high traffic hours are less suitable than areas with low traffic hours	17
Sport fishing feature count	The study region boasts many avid anglers and substantial recreational fishing opportunities. Tidal development may potentially conflict with many types of recreational fishing, thus optimal tidal siting would avoid conflict with recreational fishing.	Areas with fewer sport fishing feature counts are more suitable than those with more sport fishing feature counts.	20
Pleasure craft and yacht vessel traffic	The study area is utilized by many recreational and pleasure craft users. Potential tidal sites should avoid historically high traffic areas.	Avoid high traffic areas, most suitable locations are low traffic areas.	17
Tourism and recreation feature count	The study region attracts many tourists and recreationists who utilize the marine space in a variety of ways. Although spatial data on use intensity is lacking, along with being seasonal, this data set provides the opportunity to avoid spatial conflict with tidal development by taking tenured recreation and tourism features as an approximation of recreation and tourism uses in the study region.	Areas with fewer recreational/tourism feature counts are more suitable than areas with more recreational/tourism feature counts.	21

B.2   AHP sensitivity analysis

The final stage of the AHP entailed a sensitivity analysis, achieved by calculating the Consistency Ratio (CR) (Saaty, 1980; Stefanakou et al., 2019; Wang et al., 2009). The CR was calculated by first determining the Consistency Index (CI) from the following formula:

$$CI = \frac{\lambda_{max} - n}{n - 1} \qquad B.1$$

Where n is the number of criteria and λ_{max} is the maximum eigenvalue (Saaty, 1980). Finally, the CR is yielded by:

$$CR = \frac{CI}{RI} \qquad B.2$$

Where RI is the Random Consistency Index of a random like matrix (Table B.3) (Stefanakou et al., 2019). The value of CR should be no higher than 0.1 in order for the stakeholder's evaluation of criteria to be considered valid (Stefanakou et al., 2019).

Table B.3: Random Consistency Index of a random like matrix.

N	1	2	3	4	5	6	7	8	9	10
RI	0	0	0.58	0.90	1.12	1.24	1.32	1.41	1.45	1.49

Appendix C: Datasets and Sources

Table C.1: Overview of datasets and sources

Name	Source	Data Type	Date	Spatial reference
Tidal Resource				
1. Canadian West Coast Tidal Resource Assessment	NRCan	Vector (polygon: triangular mesh)	2014	WGS_1984_Web_Mercator_Auxiliary_Sphere
Environment				
2. BCMCA_ECO_marxan_ecol3_EXhigh_noClump	BCMCA	Vector (sampled in 2x2km grid)	2011	NAD_1983_BC_Environment_Albers
Economic				
3. BCMCA_hu_CommercialFish_GroundfishTrawl_data	BCMCA	Vector (4x4km grid)	2011	NAD_1983_BC_Environment_Albers
4. BCMCA_hu_CommercialFish_Halibut_data	BCMCA	Vector (polygon)	2012	NAD_1983_BC_Environment_Albers
5. BCMCA_hu_CommercialFish_RockfishZN_data	BCMCA	Vector (4x4km grid)	2011	NAD_1983_BC_Environment_Albers
6. BCMCA_hu_CommercialFish_Schedule2	BCMCA	Vector (4x4km grid)	2011	NAD_1983_BC_Environment_Albers
7. BCMCA_hu_CommercialFish_SalmonGillnet	BCMCA	Vector (polygon)	2010	NAD_1983_BC_Environment_Albers
8. BCMCA_hu_CommercialFish_SalmonSeine	BCMCA	Vector (polygon)	2010	NAD_1983_BC_Environment_Albers
9. BCMCA_hu_CommercialFish_SalmonTroll	BCMCA	Vector (polygon)	2010	NAD_1983_BC_Environment_Albers
10. BCMCA_hu_CommercialFish_Crab	BCMCA	Vector (4x4km grid)	2006	NAD_1983_BC_Environment_Albers
11. BCMCA_hu_CommercialFish_Geoduck	BCMCA	Vector (4x4km grid)	2011	NAD_1983_BC_Environment_Albers
12. BCMCA_hu_CommercialFish_GreenSeaUrchin	BCMCA	Vector (4x4km grid)	2011	NAD_1983_BC_Environment_Albers
13. BCMCA_hu_CommercialFish_RedSeaUrchin	BCMCA	Vector (4x4km grid)	2005	NAD_1983_BC_Environment_Albers

	Source	Format	Year	Projection
14. BCMCA_hu_CommercialFish_ Prawn	BCMCA	Vector (4x4km grid)	2006	NAD_1983_BC_Environment_Albers
15. BCMCA_hu_CommercialFish_ SeaCucumber	BCMCA	Vector (4x4km grid)	2005	NAD_1983_BC_Environment_Albers
16. BCMCA_hu_CommercialFish_ ShrimpTrawl	BCMCA	Vector (4x4km grid)	2006	NAD_1983_BC_Environment_Albers
17. BCMCA_hu_ShippingTrans_V esselTraffic	BCMCA	Vector (5x5km grid)	2010	NAD_1983_BC_Environment_Albers
18. 2018 Value of Atlantic & Pacific Coasts Commercial Landings, by Province	DFO	Excel spreadsheet	2018	N/A
19. 2018 Atlantic & Pacific Coasts Commercial Landings by Province	DFO	Excel spreadsheet	2018	N/A
Social				
20. BCMCA_hu_SportFish_Feature Count	BCMCA	Vector (2x2km grid)	2010	NAD_1983_BC_Environment_Albers
21. bcmca_hu_tourismrec_featurec ount_data	BCMCA	Vector (2x2 grid)	2010	NAD_1983_BC_Environment_Albers
Base layers				
22. MaPP_Marine_Area_EstuaryCo rrected	MaPP	Vector (polygon)	2014	BC_Albers
23. HaidaGwaii_Subregion_Oct_20 14	MaPP	Vector (polygon)	2014	BC_Albers
24. NorthCoast_Subregion_June20 15	MaPP	Vector (polygon)	2015	BC_Albers
25. NorthVancouverIsland_Subregi on_Oct3_2014	MaPP	Vector (polygon)	2014	NAD_1983_BC _Albers
26. CentralCoast_Subregion_Ver9	MaPP	Vector (polygon)	2015	BC_Albers
27. BC Basemap	GeoBC	Vector (polygon)	2014	NAD_1983_BC_Environment_Albers
28. GSR_PORTS_TERMINALS_S VW	GeoBC	Vector (point)	2019	NAD_1983_BC_Environment_Albers
29. BCMCA_ECO_Physical_Benth icClasses_DATA	BCMCA	Vector (polygon)	2010	Albers_Conical_Equal_Area

30. rced_en	NRCan	CSV	2018	N/A
Constraints				
31. TA_CRT_SVW	GeoBC	Vector (polygon)	2019	NAD_1983_BC_Environment_Albers
32. TA_PEP_SVW_polygon	GeoBC	Vector (polygon)	2018	NAD_1983_BC_Environment_Albers
33. NRCTELEMAC_elevation_elements	NRCan	Vector (polygon: triangular mesh)	2014	WGS_1984_Web_Mercator_Auxiliary_Sphere

A large portion of the datasets were taken from the British Columbia Marine Conservation Analysis (BCMCA) project This project, started in 2006, was intended to set the stage for MSP in BC by providing a set of biophysical and human use datasets to scientists, stakeholders, resource managers, and decision makers (Ban et al., 2012). 110 biophysical and 78 human use datasets were collected and processed with input from a human use working group (with two representatives from each of: commercial fisheries, recreational fisheries, ocean energy, shipping and transportation, tenures, and recreation and tourism), five ecological expert workshops, and a expert review of physical marine classification (Ban et al., 2012). The BCMCA represents a rich source of datasets that have been examined and refined by a range of stakeholders and experts with direct experience in the study region and BC more broadly, making it an invaluable resource for GIS-MCDA and other MSP work.

Marine Conservation Value sub-model

The BCMCA Ecological Marxan results Expert High (no clumps) dataset was taken to represent ecological value within the study region. This spatial analysis was undertaken to identify areas of high conservation value using available ecological data, conservation targets, and recommendations gathered from 'expert' workshops. More details on the Marxan analyses parameters and inputs can be found on the BCMCA website.

Two DFO datasets were used: seafisheries landed quantity by province, 2018; and seafisheries landed value by province, 2018. Each dataset includes a table of species and landings brought ashore by province, recorded as metric tonnes of live weight and thousands of dollars, respectively. This data has been collected annually since 1990.

The BCMCA data provides information for singular or grouped subspecies in individual GIS shapefiles showing spatial distribution, total catch in each fishing zone, along with the duration of the data collection. These data sets were screened and sorted, removing those fisheries that have been closed in the study region (e.g. herring), or do not have corresponding or comparable species data with that available from the DFO (e.g. sablefish). In total 14 data sets were used encompassing approximately 18 species.

Table C.2: Department of Fisheries and Oceans Canada species landing value and volume data.

	Species (BCMCA)	Years of Data Collection	Species (DFO)	Total catch values (k$)	Total catch volume (metric tonnes)	Price ($/kg)
			Groundfish			
1	Groundfish Trawl	1996-2004	Ground	175,269	129,645	1.35
2	Rockfish hook and line	1993-2004	Ground	175,269	129,645	1.35
6	Halibut	1991-2010	Halibut	61,748	3,812	16.1
7	Schedule II	1996-2004	Cod	980	577	1.7 Average 0.91
			Dogfish	11	100	0.11 Average 0.91

Salmon						
11	Gillnet	2001-2007	Salmon (all)	46,255	12,893	3.59
12	Seine	2001-2007	Salmon (all)	46,255	12,893	3.59
13	Troll	2001-2007	Salmon (all)	46,255	12,893	3.59
Shellfish						
14	Dungeness Crab	2000-2004	Crab, other	51,059	3,802	13.42
15	Geoduck	2000-2005	Clams/quahaug	52, 171	1,986	26.27
16	Green Sea Urchin	2000-2005	Sea urchin	7,398	3,126	2.37
17	Red Sea Urchin	2000-2005	Sea urchin	7,398	3,126	2.37
18	Prawn	2001-2004	Shrimp	23,866	1,411	16.9
19	Sea Cucumber	2000-2005	Sea cucumbers	8,609	1,736	4.59
20	Shrimp	1996-2004	Shrimp	23,866	1,411	16.9

Appendix D: Diesel Fuel Price for Current Suitability and Tidal Stream Turbine Device Specifications

D.1 Diesel fuel price for current suitability distribution

While the average diesel price for Victoria from January 2018-April 2020 was $1.319/L, the Organization of the Petroleum Exporting Countries price war between Russia and Saudi Arabia, along with the global impacts of COVID-19 depressing demand and reducing prices, highlights the volatility of fossil fuel prices while also not being indicative of 'business as usual' average prices (NRCan, 2020a; Valleau, 2020). Thus, diesel price data from January-April 2020 was omitted yielding an average price of $1.352/L. Fuel for use by First Nation's or for sale on First Nations reserves is exempt from taxes, while BC Hydro could not be reached at the time to confirm whether taxes are paid on fuel within communities serviced by their Non-Integrated Areas (NIA) program. Thus, this study assumed the communities assessed are tax exempt, removing the average $0.40/L in taxes and yielding a price of $0.95/L. There are also shipping costs associated with the transport of diesel to remote communities, estimated at $0.10/L in this study, similar to other studies on the BC coast (Lafleur et al., 2020). This yielded a fuel price for input into Truelove's equations of $1.05/L

D.2 TST device specifications

Table D.1: Tidal Stream Turbine Device Specifications.

Turbine performance	Cut out speed (m/s) Bolded if taken from online, otherwise calculated based on Equation 4.3.	Rated water velocity (m/s)	Rated power (kW)	Source
Guinard P66	4.95	3	3.5	https://www.guinard-energies.bzh/wp-content/uploads/Flyer-P66-Guinard-Energies.pdf
Guinard P154	5.94	3.6	20	https://www.guinard-energies.bzh/en/our-products/p154-hydrokinetic-turbine/
Schottel 54	**4.6**	2.6	54	http://www.blackrocktidalpower.com/fileadmin/data_BRTP/pdf/STG-datasheet.pdf
Schottel 62	**6**	3	62	http://www.blackrocktidalpower.com/fileadmin/data_BRTP/pdf/STG-datasheet.pdf
Schottel 70	**6.75**	3.8	70	http://www.blackrocktidalpower.com/fileadmin/data_BRTP/pdf/STG-datasheet.pdf
SeaGen	4.125	2.5	1200	file:///C:/Users/riley/AppData/Local/Packages/Microsoft.MicrosoftEdge_8wekyb3d8bbwe/TempState/Downloads/Power_Quality_Performance_of_the_Tidal_Energy_Conv%20(1).pdf Power quality performance of the tidal energy converter SeaGen
SMART Free Stream	5.115	3.1	5	https://www.smart-hydro.de/wp-content/uploads/2015/12/Datasheet_SMART_Freestream.pdf
SMART Monofloat	4.62	2.8	5	https://www.smart-hydro.de/wp-content/uploads/2015/12/Datasheet_SMART_Monofloat.pdf

Appendix E: Layer Histograms

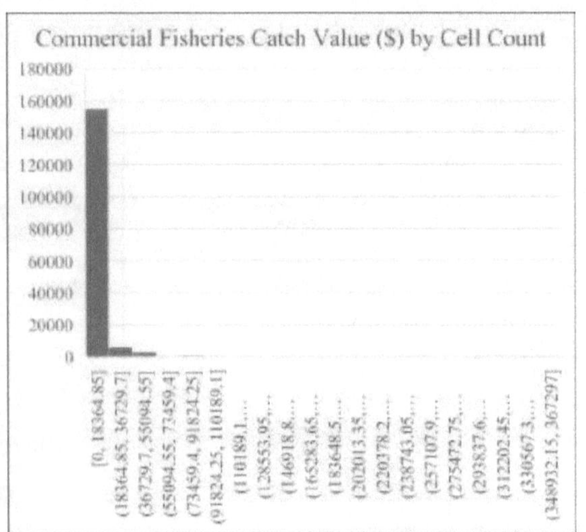

Figure E.1: Commercial fisheries catch value distribution.

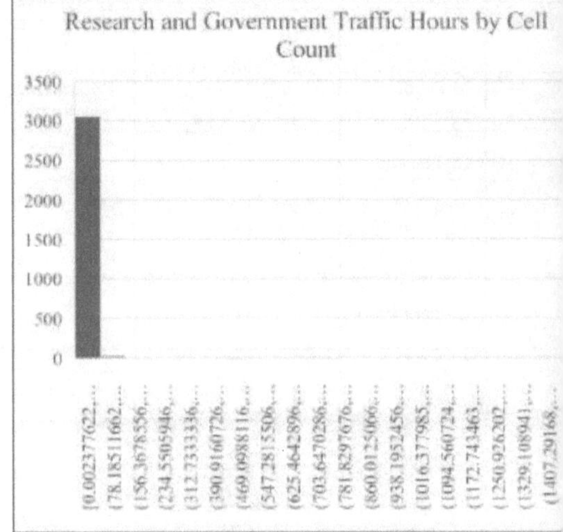

Figure E.2: Research and government vessel traffic hours distribution.

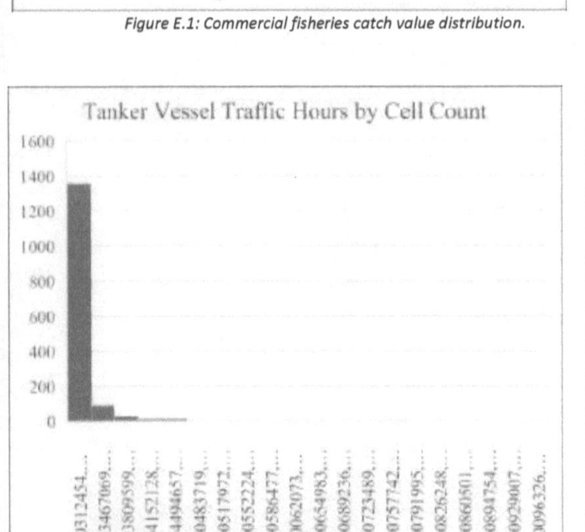

Figure E.3: Tanker vessel traffic hours distribution.

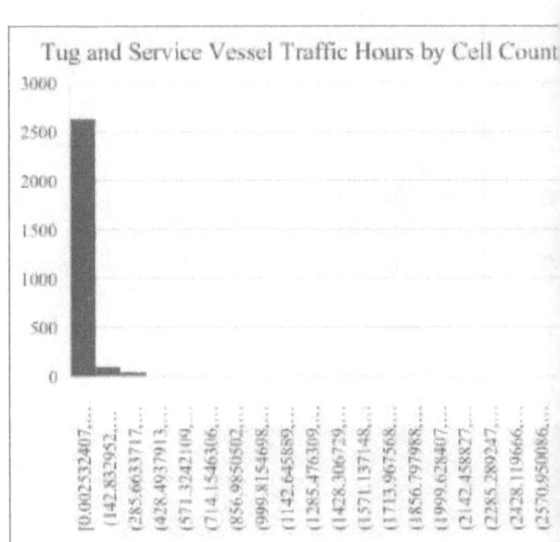

Figure E.4: Tug and Service vessel traffic hours distribution.

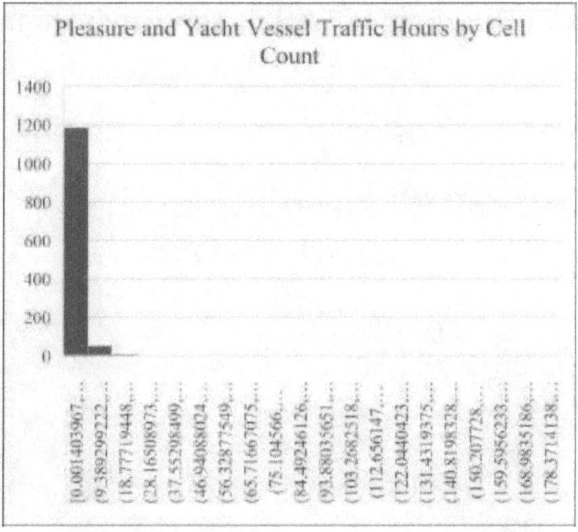

Figure E.5: Pleasure and Yacht vessel traffic hours distribution.

Appendix F: Economic, Social, and Environmental Use Layers

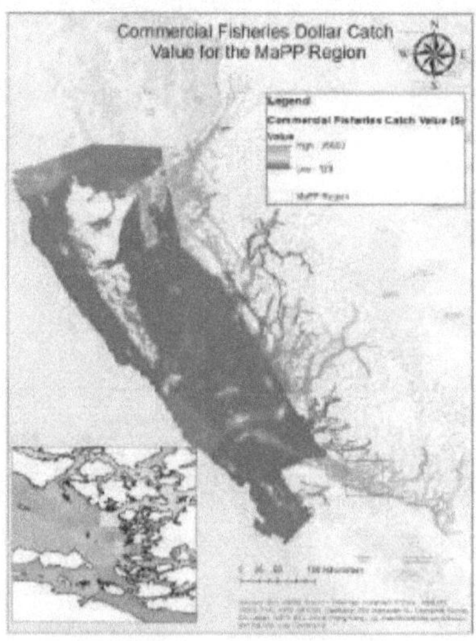

Figure F.1: Commercial Fisheries Catch Value.

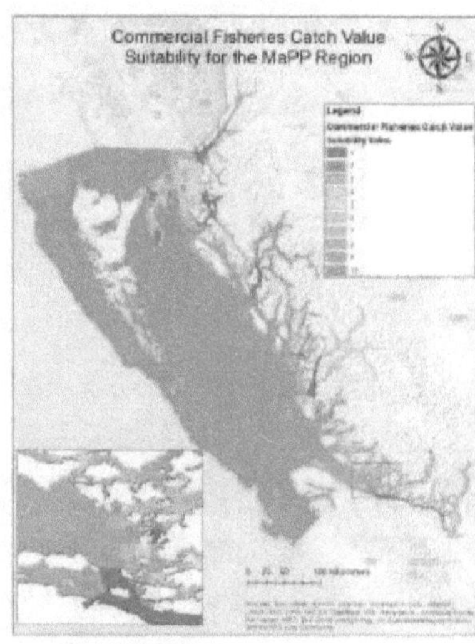

Figure F.2: Commercial Fisheries Catch Value suitability.

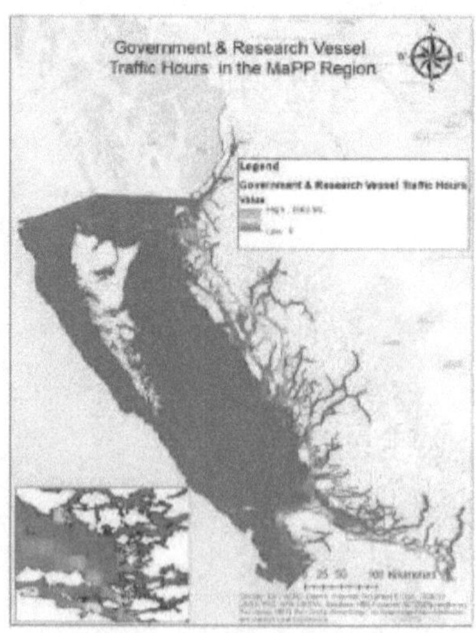

Figure F.3: Government and Research vessel traffic.

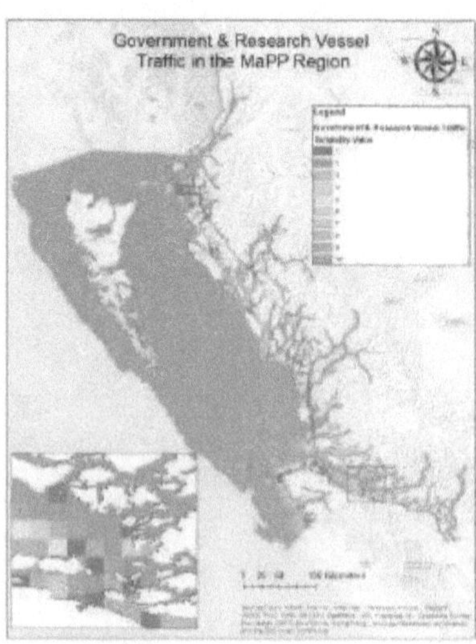

Figure F.4: Government and research vessel traffic suitability.

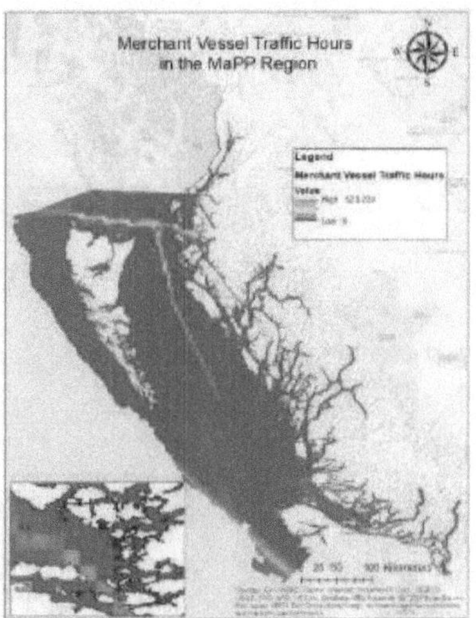

Figure F.5: Merchant vessel traffic.

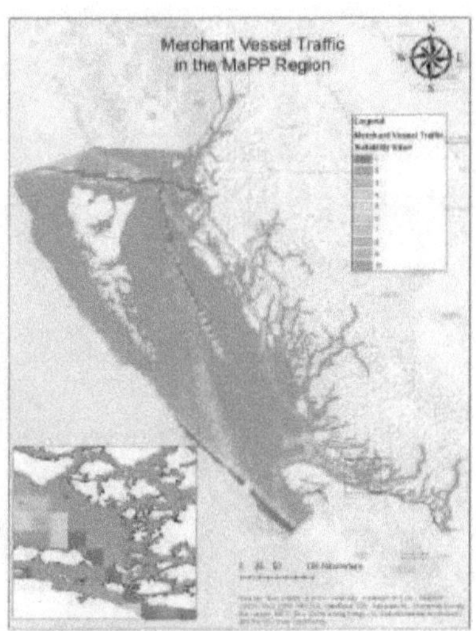

Figure F.6: Merchant vessel traffic suitability.

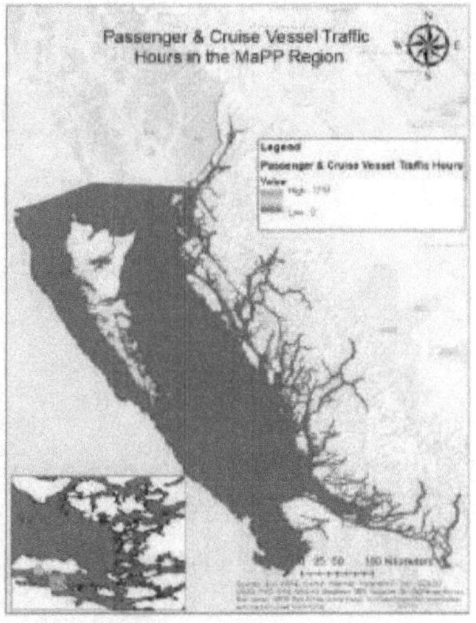

Figure F.7: Passenger and Cruise vessel traffic.

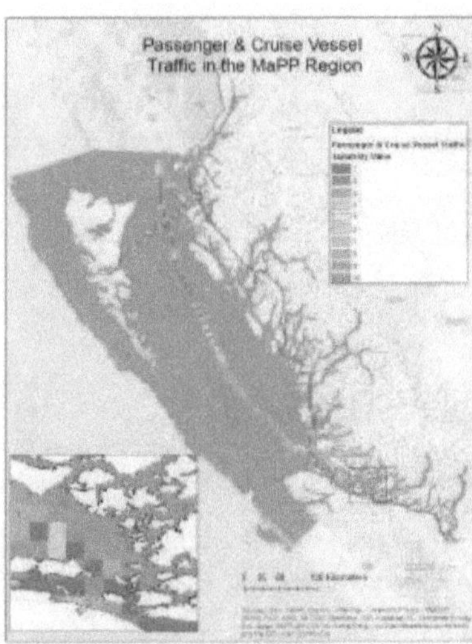

Figure F.8: Passenger and Cruise vessel traffic suitability.

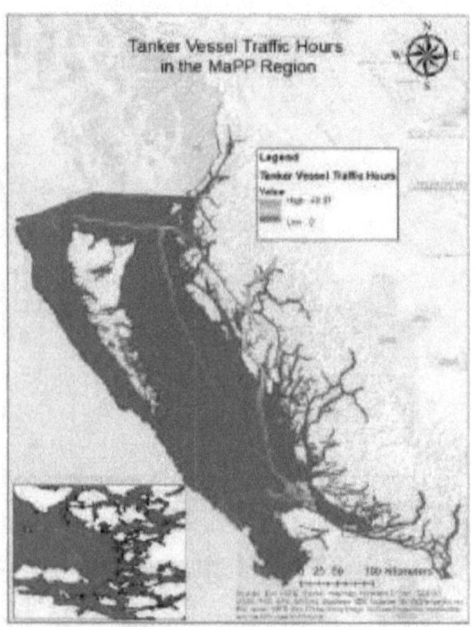

Figure F.9: Tanker vessel traffic.

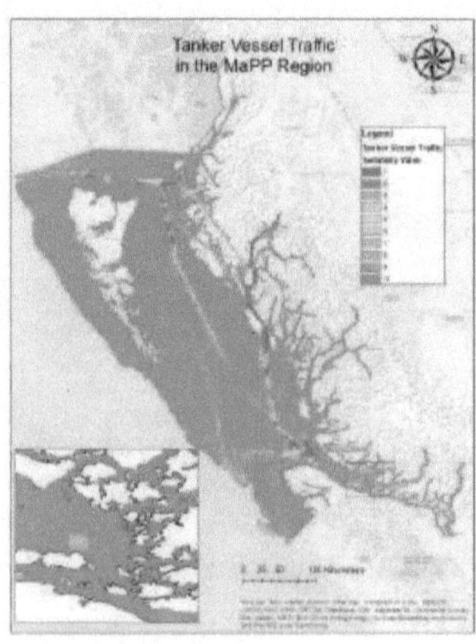

Figure F.10: Tanker vessel traffic suitability.

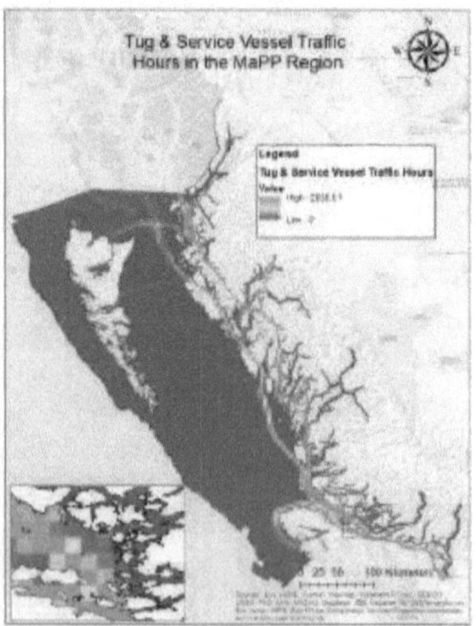

Figure F.11: Tug and Service vessel traffic.

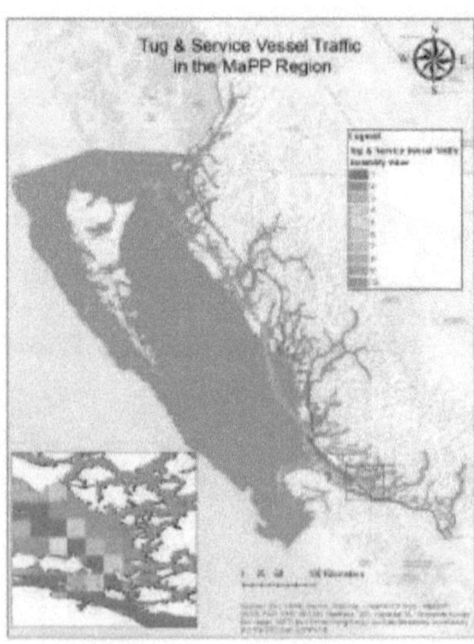

Figure F.12: Tug and Service vessel traffic suitability.

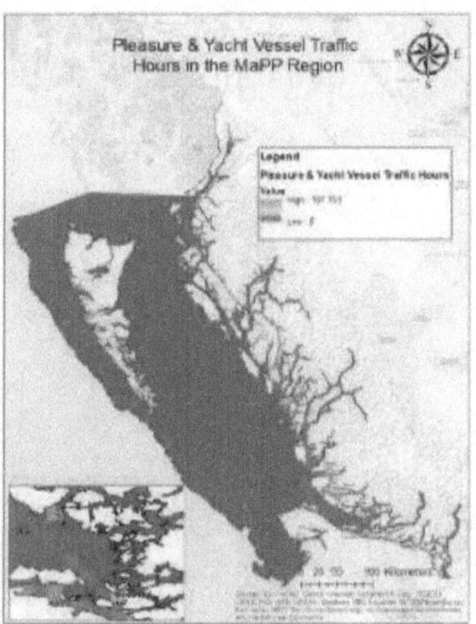

Figure F.13: Pleasure and Yacht traffic.

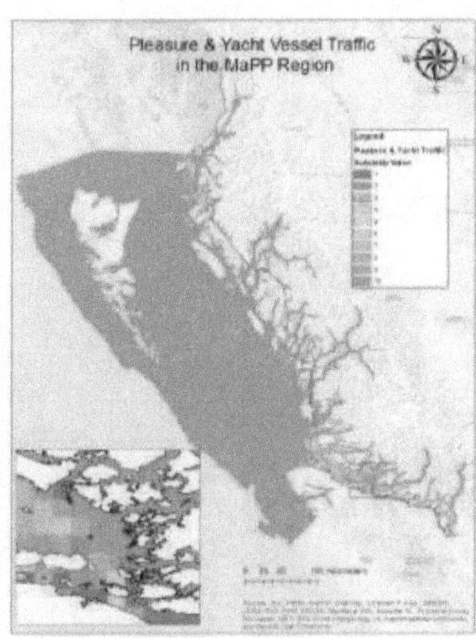

Figure F.14: Pleasure and Yacht traffic suitability.

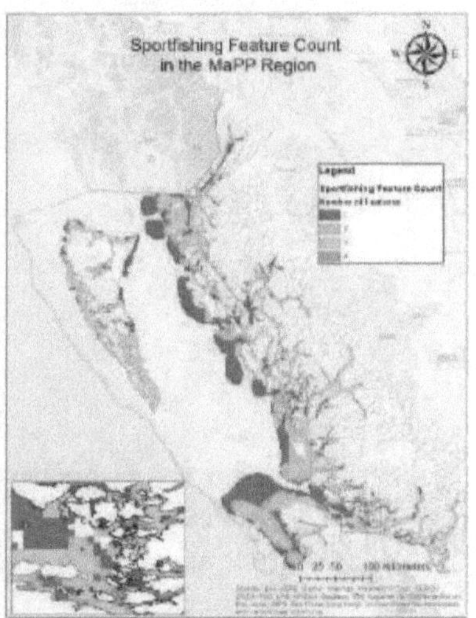

Figure F.15: Sportfishing feature count.

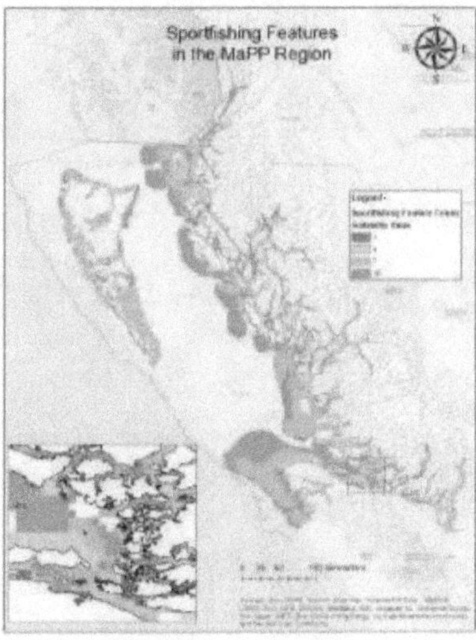

Figure F.16: Sportfishing feature count suitability.

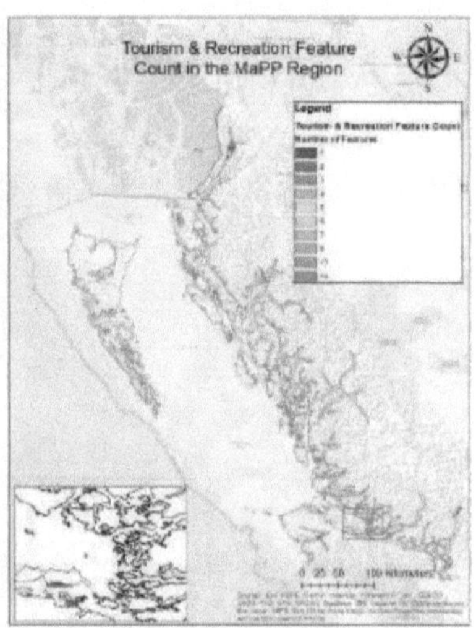

Figure F.17: Tourism and Recreation feature count.

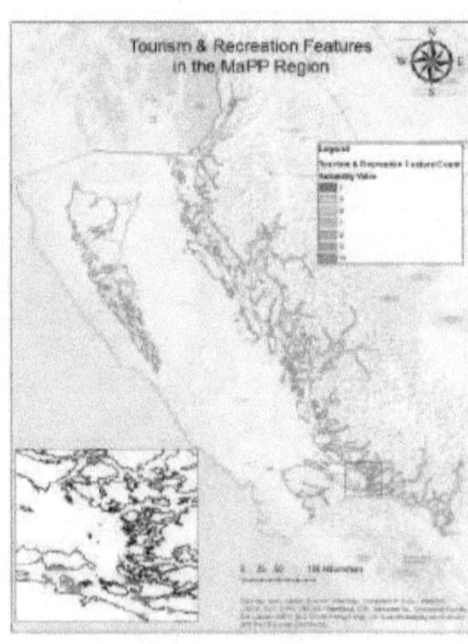

Figure F.18: Tourism and Recreation feature count suitability.

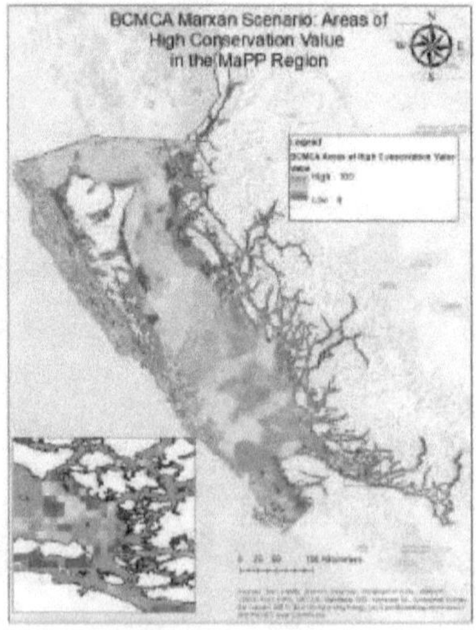

Figure F.19: High Conservation value areas

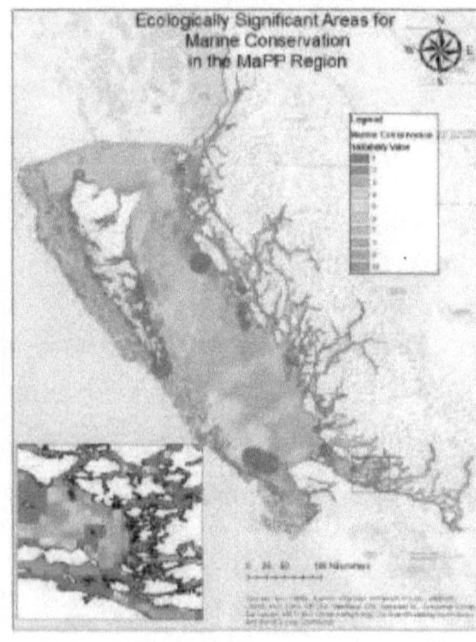

Figure F.20: High Conservation value areas suitability.

Appendix G: AHP Matrices and Sub-model Sensitivity Analyses

G.1   AHP Matrices for Economic and Social uses

G.1.1: AHP matrix for Economic Uses provided to participants for question 5.

	Commercial fisheries catch value	Government/ research vessel traffic	Merchant vessel traffic	Passenger and cruise vessel traffic	Tanker vessel traffic	Tug and service vessel traffic
Commercial fisheries catch value	1					
Government/ research vessel traffic		1				
Merchant vessel traffic			1			
Passenger and cruise vessel traffic				1		
Tanker vessel traffic					1	
Tug and service vessel traffic						1

G.1.2: AHP matrix for Social Uses provided to participants for question 6.

	Sport fishing feature count	Pleasure craft and yacht traffic	Recreation and tourism feature count
Sport fishing feature count	1		
Pleasure craft and yacht traffic		1	
Recreation and tourism feature count			1

G.2   Sub-model sensitivity analysis

A univariate sensitivity analysis for the entire MaPP region was used to assess the Economic and Social Uses Suitability sub-model criteria weightings. For this, an equal weights scenario was used as a reference sub-model, along with 'focus' scenarios for each criterion. For both the Economic and Social Uses equal weight scenario, the highest ranked criteria by participants received 'extra' weightings as the *Weighted Overlay* tool does not allow decimal values (e.g. 17 for the four highest economic uses rather than all being 16.66, and 34 for the highest ranked social use rather than all being 33.33). Each of the other scenarios was focused on altering the weight of one criterion, while the others were equally weighted. The single criterion in question was allocated a weight greater-or-equal to twice the weight of one of the other criteria.

Table G.2.1: Economic Uses sub-model sensitivity analysis weightings.

Scenario	Commercial fisheries catch value	Government / research vessel traffic	Merchant vessel traffic	Passenger and cruise vessel traffic	Tanker vessel traffic	Tug and service vessel traffic
Equal weights	17	17	16	17	16	17
CFCV focus	30	14	14	14	14	14
Government focus	14	30	14	14	14	14
Merchant focus	14	14	30	14	14	14
Passenger focus	14	14	14	30	14	14
Tanker focus	14	14	14	14	30	14
Tug and service focus	14	14	14	14	14	30

Table G.2.2: Economic Uses sub-model percent area covered for Equal Weights (EW) scenario and percent change for each focus scenario relative to EW.

Value	Equal Weight	CFCV focus	Government focus	Merchant focus	Passenger focus	Tanker focus	Tug and service focus
1	0.01%	0.00%	0.00%	0.00%	0.00%	0.00%	0.00%
2	0.27%	0.00%	0.02%	0.00%	0.06%	-0.16%	0.05%
3	0.44%	0.19%	0.38%	0.02%	0.34%	-0.23%	0.29%
4	0.91%	-0.48%	-0.15%	-0.07%	-0.08%	0.02%	-0.23%
5	1.31%	-0.16%	-0.08%	0.57%	0.61%	0.43%	0.15%
6	2.27%	-0.23%	-0.08%	0.05%	-1.02%	0.27%	-0.77%
7	2.43%	1.33%	0.49%	1.65%	0.12%	0.09%	0.19%
8	6.81%	-1.26%	-2.30%	-1.83%	-1.58%	-1.67%	-2.08%

Table G.2.3: Social Uses sub-model sensitivity analysis weightings.

Scenario	Tourism and recreation feature count	Sportfishing feature count	Pleasure craft and yacht traffic
Equal weights	34	33	33
Recreation and tourism focus	50	25	25
Sportfishing focus	25	50	25
Pleasure craft and yacht focus	25	25	50

Table G.2.4: Social uses sub-model EW percent area covered by suitability value and percent change for each focus scenario relative to EW.

Value	Equal Weight	Tourism and recreation focus	Sportfishing focus	Pleasure craft and yacht focus
1	0.00%	0.00%	0.00%	0.00%
2	0.00%	0.00%	0.00%	0.00%
3	0.06%	-0.06%	0.03%	0.02%
4	0.26%	-0.26%	-0.02%	0.52%
5	0.63%	-0.31%	-0.24%	0.53%
6	1.62%	-0.45%	-0.10%	-0.97%
7	3.27%	-1.52%	-0.08%	-2.52%
8	14.27%	2.57%	0.02%	-10.40%
9	79.79%	-0.01%	-12.17%	0.21%
10	0.10%	0.05%	12.56%	12.61%

Appendix H: Phase I Model Sensitivity Analysis

Table H.1: North Vancouver Island sub-region percent area covered for the equal weight's scenario and percent change relative to equal weights for the NVI Tidal development perspective scenario.

Value	Equal Weight	Tidal development perspective	Change in percent coverage (TDP-EW)
0	32.92%	32.92%	0.00%
1	0.00%	0.00%	0.00%
2	0.78%	6.01%	5.22%
3	8.42%	15.83%	7.41%
4	15.58%	14.21%	-1.38%
5	15.26%	15.79%	0.53%
6	16.69%	10.65%	-6.04%
7	8.93%	3.65%	-5.28%
8	1.40%	0.93%	-0.47%
9	0.01%	0.00%	-0.01%
10	0.00%	0.00%	0.00%

Table H.2: Central Coast sub-region percent area covered for the equal weight's scenario and percent change relative to equal weights for the CC Tidal development perspective scenario.

Value	Equal Weight	Tidal development perspective	Change in percent coverage (TDP-EW)
0	24.22%	24.22%	0.00%
1	0.00%	0.00%	0.00%
2	0.00%	0.00%	0.00%
3	0.00%	5.38%	5.38%
4	5.38%	1.97%	-3.41%
5	1.88%	0.09%	-1.79%
6	0.27%	58.65%	58.39%
7	66.46%	9.69%	-56.77%
8	1.79%	0.00%	-1.79%
9	0.00%	0.00%	0.00%
10	0.00%	0.00%	0.00%

Table H.3: North Coast sub-region percent area covered for the equal weight's scenario and percent change relative to equal weights for the NC Tidal development perspective scenario.

Value	Equal Weight	Tidal development perspective	Change in percent coverage (TDP-EW)
0	58.79%	58.79%	0.00%
1	0.00%	0.00%	0.00%
2	0.00%	0.00%	0.00%
3	0.00%	2.53%	2.53%
4	3.21%	17.95%	14.74%
5	20.73%	14.62%	-6.11%
6	14.81%	4.75%	-10.06%
7	2.47%	1.36%	-1.11%
8	0.00%	0.00%	0.00%
9	0.00%	0.00%	0.00%
10	0.00%	0.00%	0.00%

Table H.4: Haida Gwaii sub-region percent area covered for the equal weight's scenario and percent change relative to equal weights for the HG Tidal development perspective scenario.

Value	Equal Weight	Tidal development perspective	Change in percent coverage (TDP-EW)
0	54.10%	54.10%	0.00%
1	0.00%	0.00%	0.00%
2	0.00%	0.00%	0.00%
3	0.00%	0.00%	0.00%
4	0.19%	13.89%	13.71%
5	14.57%	16.05%	1.47%
6	17.22%	10.99%	-6.23%
7	7.60%	4.97%	-2.62%
8	6.33%	0.00%	-6.33%
9	0.00%	0.00%	0.00%
10	0.00%	0.00%	0.00%

Appendix I: Phase I GIS-MCDA Sub-model Results

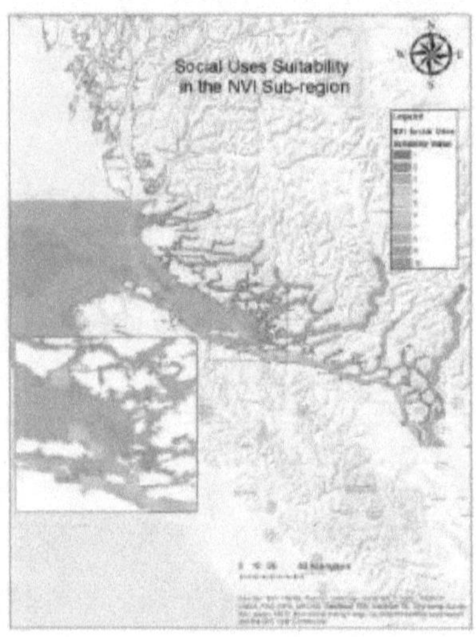

Figure I.1: Social Uses suitability in the NVI sub-region.

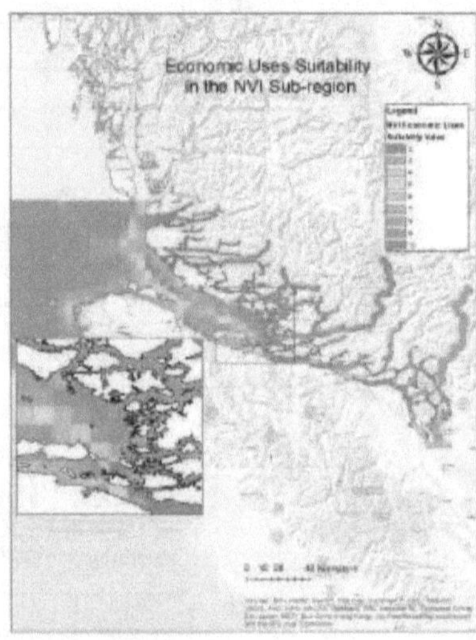

Figure I.2: Economic Uses suitability in the NVI sub-region.

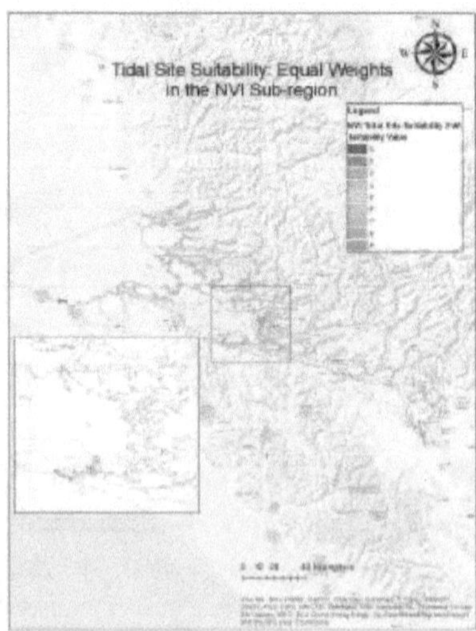

Figure I.3: EW tidal site suitability in the NVI sub-region

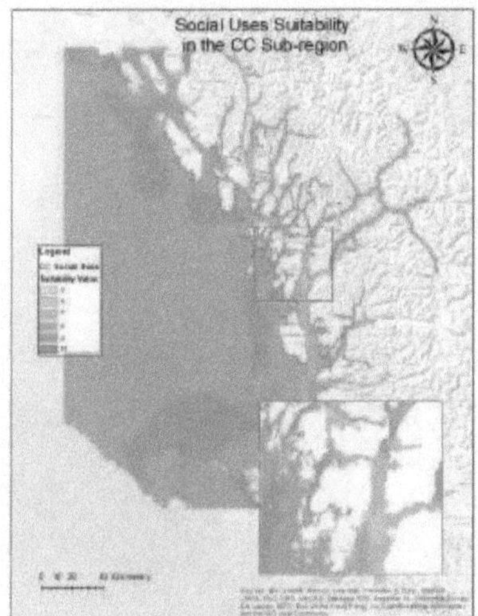

Figure I.4: Social Uses suitability in the CC sub-region.

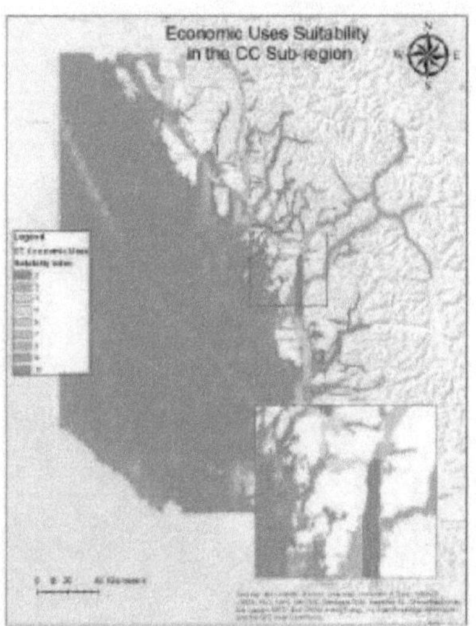

Figure I.5: Economic Uses suitability in the CC sub-region.

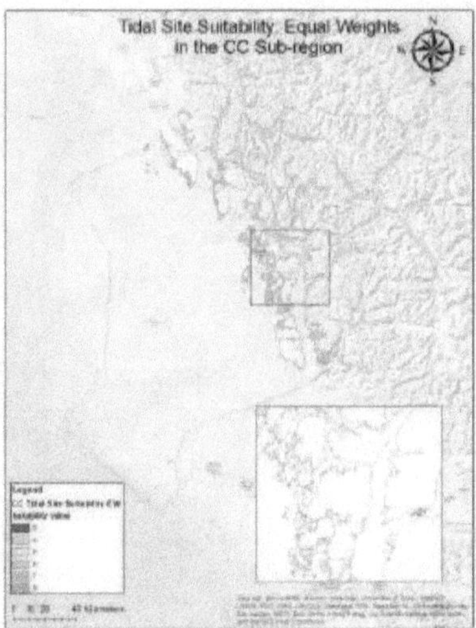

Figure I.6: Tidal site suitability values for the Equal Weights scenario in the CC sub-region.

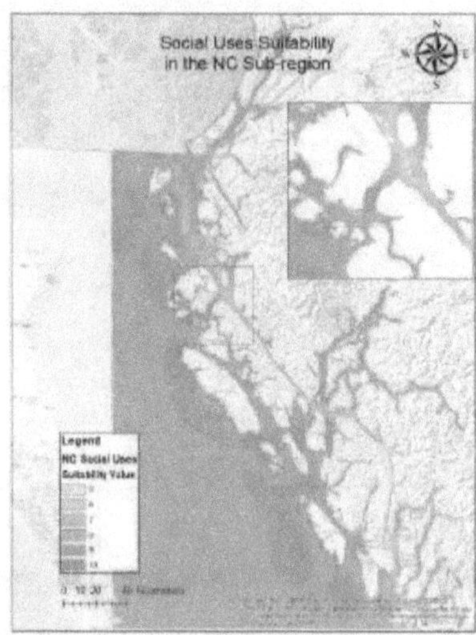

Figure I.7: Social Uses suitability in the NC sub-region.

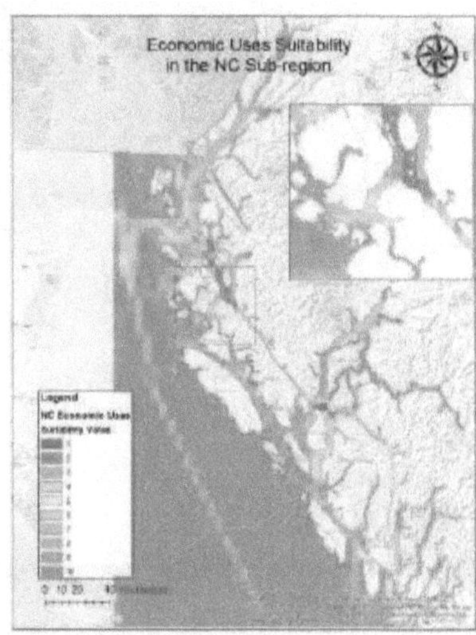

Figure I.8: Economic Uses suitability in the NC sub-region.

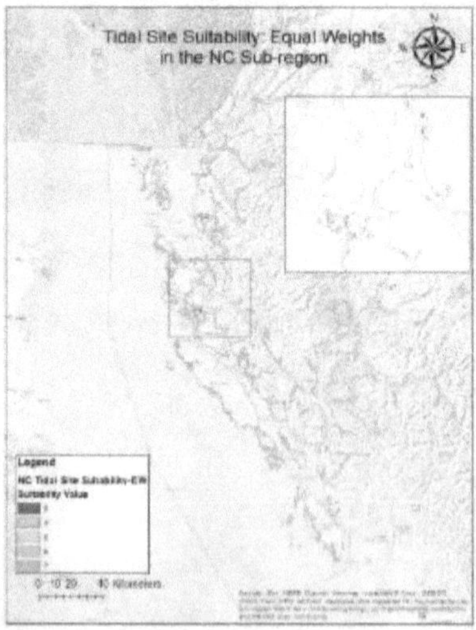

Figure I.9: Tidal site suitability value for the Equal Weights scenario in the NC sub-region.

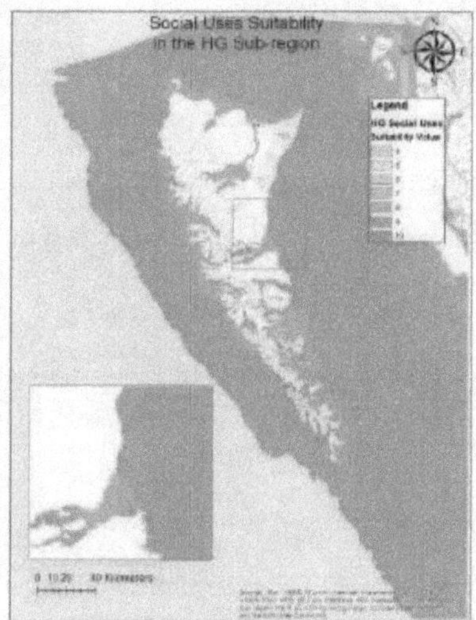

Figure I.10: Social Uses suitability in the HG sub-region

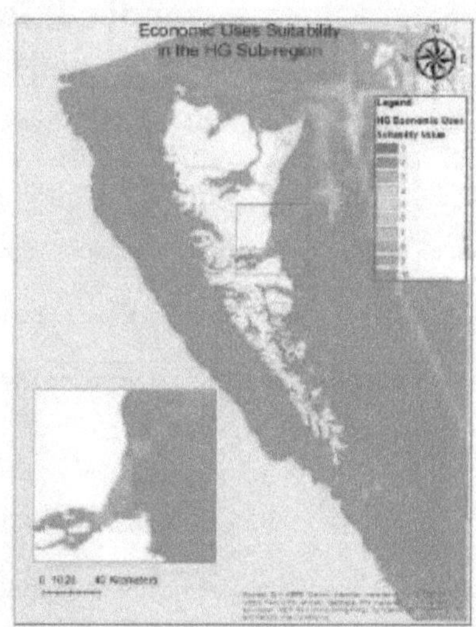

Figure I.11: Economic Uses suitability in the HG sub-region.

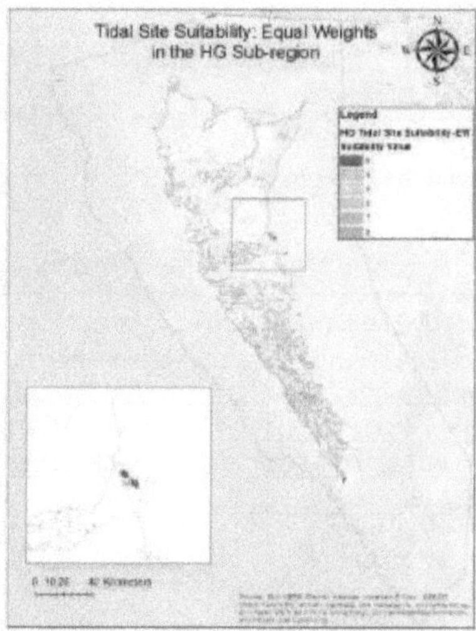

Figure I.12: Tidal site suitability value for the Equal Weights scenario in the HG sub-region.

Appendix J: Port Selection Rationale

The ports data was screened to select DESCRIPTN = Terminal, which are defined as being "…generally associated with commercial, sea going operations", with the TYPE_USE equal to: Vessel Services, Passenger Cruise Vessel Services, Not in use, Fishing, or Energy (N.S. DEM, 2011). These use types were selected as they either indicated the provision of vessel services, an unused area, or a perceived compatibility with the requirements of MRE deployment.

Of the identified terminals, Gold River and Victoria were removed as they were beyond the threshold distance of 400km from sites, along with being located on Vancouver Island. This would imply an additional travel distance from the mainland (e.g. it would logically be more cost efficient to set out from a continental port). Furthermore, the Kitimat Shell facility and Richmond facility were removed as both are occupied by existing activities, with the Kitimat Eurocan facility removed as it is not in use/was purposed for fossil fuel exportation. Thus two ports were chosen: the Port of Vancouver and the Port of Prince Rupert.

Table J.1: Port selection rationale (N.S. DEM, 2011).

Port requirements	
Number of berths	one to two
length of wharf	100-150m
Water depth	>4m below low normal tide
Back up land/staging area	4-5hectares
Crane capacity	60+ tonnes

Appendix K: Phase II Equal Weights Model Results

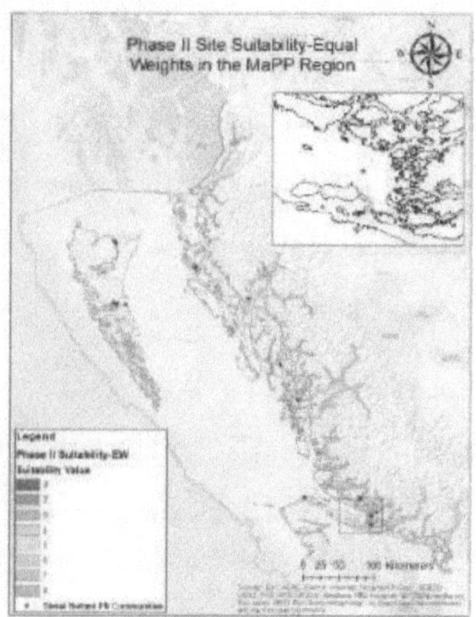

Figure K.1: MaPP region Equal Weights Phase II site suitability.

Appendix L: Tidal Stream Turbine Levelized Cost of Energy Inputs

Project lifetime

For other LCOE case studies, project lifetime ranged from 20 years (Nissen & Harfst, 2019; Segura et al., 2017a; Vazquez & Iglesias, 2016) to 25 years (Bricker et al., 2017). Meygen 1A also has a project lifetime of 25 years (Black & Veatch, 2020). Based on existing studies and the approximate lifetime of BC Hydro's diesel stations a project lifetime of 20 years was used.

Project start

The start year for the project is taken as 2023, which is both inline with existing funding opportunities and policy goals, while also being a an ambitious but feasible small-scale project timeline assuming a community applies for funding in 2020-2021 (Coast Funds, 2019, Province of British Columbia, 2018).

Inflation rate

The inflation rate is a critical component of any economic analysis. Inflation in Canada since records began (1914) to 2020 was 3.01%, while the most recent 30-year average (1990-2020) was 1.88% (Bank of Canada, 2020). This study assumes a future inflation rate of 1.88%

Canadian-US exchange rate

As many of the cost metrics feeding into the CAPEX and OPEX formulas are in US $ a conversion rate is required, taken as the average exchange rate over the past three years (2017-2019) with an average value of $1CAD = $0.76US (Bank of Canada, 2020).

Discount rate

Discount rate is a one of the most critical inputs into economic analyses, with tidal energy LCOE's being no different (Dalton et al., 2015). Discount rates used in TST studies vary widely, ranging from 3-15% and can be found in Table N.1.

Table N.1: Discount rates used in existing TST studies.

Author(s)	Discount rate(s) used
Oxera, 2011	Estimated a low rate of 11% and high rate of 15% to be used by 2020
OES, 2015	10%
Stothers & Klaptocz, 2016	12%
Vazquez & Iglesias, 2016c	Low rate of 6% and high rate of 12%
Bricker et al., 2017	3%, 5%, 10%
Mavi Innovations, 2017	14%
Smart & Noonan, 2018	10%
Shelley, 2019	11%

While discount rates reflect risk and uncertainty, there are often social discount rates applied by government towards projects which enhance or provide social benefits. Based on the studies assessed, a discount rate of 10% is used to reflect broader economic sentiment in terms of investment risk (i.e. private-FN partnership) and a social discount rate of 4% is used to represent subsidized or socially beneficial analysis within the tidal cost calculator.

Parameters not included

Several parameters were not included, such as Feed in tariffs for which the offer program in BC ended in 2012 and the escalation rate was set as zero.

www.ingramcontent.com/pod-product-compliance
Lightning Source LLC
LaVergne TN
LVHW042107190726
843493LV00006B/1388